The Racial Justice Series
By
Roberto Schiraldi

Healing Love Poems
for white supremacy culture:
Living Our Values

Unexpurgated*Racial Justice Poetry
with Healing Meditations

Men and Racism:
The Healing Path

Men and Racism: The Healing Path

A Courageous and Compassionate Journey Through The Fear Of Being Vulnerable

Francesco Roberto
Vincenzo Schiraldi

BALBOA.PRESS
A DIVISION OF HAY HOUSE

Copyright © 2024 Roberto Schiraldi.

All rights reserved. No part of this book may be used or reproduced by any means, graphic, electronic, or mechanical, including photocopying, recording, taping or by any information storage retrieval system without the written permission of the author except in the case of brief quotations embodied in critical articles and reviews.

Balboa Press books may be ordered through booksellers or by contacting:

Balboa Press
A Division of Hay House
1663 Liberty Drive
Bloomington, IN 47403
www.balboapress.com
844-682-1282

Because of the dynamic nature of the Internet, any web addresses or links contained in this book may have changed since publication and may no longer be valid. The views expressed in this work are solely those of the author and do not necessarily reflect the views of the publisher, and the publisher hereby disclaims any responsibility for them.

The author of this book does not dispense medical advice or prescribe the use of any technique as a form of treatment for physical, emotional, or medical problems without the advice of a physician, either directly or indirectly. The intent of the author is only to offer information of a general nature to help you in your quest for emotional and spiritual well-being. In the event you use any of the information in this book for yourself, which is your constitutional right, the author and the publisher assume no responsibility for your actions.

Any people depicted in stock imagery provided by Getty Images are models, and such images are being used for illustrative purposes only. Certain stock imagery © Getty Images.

Print information available on the last page.

ISBN: 979-8-7652-5412-7 (sc)
ISBN: 979-8-7652-5411-0 (hc)
ISBN: 979-8-7652-5410-3 (e)

Library of Congress Control Number: 2024915441

Balboa Press rev. date: 10/11/2024

CONTENTS

Cover Picture ix
Preludes xi
Dedication xiii
Gratitude xv

Part I   The Heart Of Our Roots
Rumi Poem 1
Letter 1 2
Vulnerability and Strength 5
Mitakuye Oyasin 11
Letter 2 17
Fear of Tears …. or……Honoring Our
Tears….A Truly Radical Path to Healing ….. 19
Macho Man….. from boyhood to man 26
Men and Racism, The Compassionate
Path to Healing 29
white supremacy culture 33
On Being Vulnerable 37
Vulnerability 41
A Synthesis and Re-Imagining 45
Values 48
Overview: Racism/Men/Vulnerable, Initial Guide 52
Native 60
Stories of Fear, Sacred, Love, Vulnerability 62

We Men 75
Boys To Men: Beyond Diversity 77
My Sacred Feminine 87
Mental health and wealthy white hetero male
supremacy, the Dilemma... 89
Monkey Suit...Defined 93

Part II   Living Change
Men for Racial Healing and Justice (proposal)............... 97
Maximum Security 101
Double Jeopardy (Veterans of Color)........................... 102
Dog Soldier 109
Men Stepping Up (Veterans For Peace proposal) 111
Climate Crisis and Militarism..118
Responsible 120
The Web Of Life 120
Coming Together..121
Poem of Gratitude and Hope... 122
Revitalizing the Land... 123
The Earth Is Our Mother.. 125
Recommendations for Police Reform (proposal) 126
Truth and Conciliation 137
Truth and Conciliation, Princeton (proposal)................ 139
Trust 160
The Sacred, Back to the Beginning 163
All Life Is Sacred (proposal)... 164
Letter to the President 182

Outline for Core Curriculum Course (proposal)	186
I Have A Dream	213
Two-Spirit	216
The bully Balance	219
Lessons From Dad	228
Assimilation Nation	230
USA...USA	242
white supremacy as Addiction / 12 Step Recovery	244
What If Washington Was Gay?	247
One Brave Man	250
Gay/Straight Continuum	256
Initial Strategies For Engaging In Racial Justice Work	257
Rosa Parks	261
A Final Racial Healing Story	263
The Children	265
Racial Justice Blessing	269
Healing Meditation	271
Two Final Eileen Stories	273
Epilogues	277
Dear Kindred	279
Gratitude To The Max	280
Gratitude Prayer	281
Additional Recommendations	282
Author Info	284

Cover Picture

I was hesitant to use this picture …..…
because of the railings...

...less free...

And then I re-membered.......

…...this is a beautiful....

…...yet difficult journey......

…..... and we can all use all the support we can get...

..thus the railings.

Hope you can feel the loving intent.

Preludes

"You can't hold a man down, without staying down with him."

Booker T. Washington

"Until the killing of black men, black mothers' sons, becomes as important to the rest of the country, as the killing of a white mother's son, we who believe in freedom cannot rest."

Ella Baker

Nothing is so strong as gentleness.
Nothing so gentle as real strength.

Lakota Saying.

Fear is the root of hatred.
Love is the antidote to both.

Traditional Wisdom

Dedication

To All Good Men of Good Hearts
    And Good Will.

To All Good People of Good Hearts
    And Good Will

To You For Choosing to Open
    This Book

To Everyone...as We Are All Related
    All Relatives, All One

Gratitude

To The Love Of My Life Eileen who passed five months ago.

You chose, supported, encouraged, and held me in your loving heart and arms as we traveled on this beautiful road through Fear, Sacred, Love, Vulnerability.

I Breathe...and I Feel You Inside me. I see you high in the sky....floating gracefully with the wind.

Part I

The Heart Of Our Roots....

With Gentleness and Strength

This poem by Rumi, is one of my favorites, and a great help for us with our feelings:

THE GUEST HOUSE

This being human is a guest house.
Every morning a new arrival.

A joy, a depression, a meanness,
some momentary awareness comes
as an unexpected visitor.

Welcome and entertain them all!
Even if they are a crowd of sorrows,
who violently sweep your house
empty of its furniture,
still, treat each guest honorably.
She may be clearing you out
for some new delight.

The dark thought, the shame, the malice.
meet them at the door laughing and invite them in.

Be grateful for whatever comes.
because each has been sent
as a guide from beyond.

-Jelaluddin Rumi

Letter 1.

A Love Letter to All My Brothers:

My Dear Brothers...... I believe we are all connected...all related.....All brothers,

Please know that what follows here is my humble effort to bring us all together.....to address and heal this disease of racism which so hurts and plagues .....us ALL.

<u>I'm writing to all you men of good hearts and good will.....because I have faith that if we all choose to come together ....and lead with our hearts..... with a little....make that.. a lot....of support and leadership from Indigenous, Black, Latinx, Asian Pacific Islanders, Women, Gay folks, Trans folks, Immigrants/Refugees....all of us........we can co-create a world that truly serves us ALL.</u>

Why do I title this book 'Men and Racism'? Certainly not to attack, belittle, disrespect, or guilt trip us men...... as I am part of "us men and racism"...and I don't have any desire to put myself or anyone else down........And.....I know that we men....and when I say "we"...it's because I feel deep in my soul....that to heal racism and all the other woes of this world.....we each have to own and acknowledge our part of it.....lest we continue to struggle, seeing each other as threats to our tenuous hold on our pseudo security / material belongings. Thus "Our Fear Of Being Vulnerable", in a world that often doesn't feel safe for us and our loved ones.

And why 'The Courageous and Compassionate Path to Healing'? Because this can be really difficult, painful work, which will take all we have to give.....together. We need each other. Facing fear takes so much courage and compassion....to finally get to the place of loving healing.

Pointing fingers.....attacking others....doesn't work....and only creates defensiveness. Looking at ourselves in the mirror, with impeccable integrity, courage, humility, compassion can lead us all to freedom. We all are acting in ways that reflect our backgrounds...our core training about the values to live by....of what it means to be male... whether they be healthy, caring, compassionate values.... or more selfish, fear based, defensive posturing to protect what we have. There have been various men's movement efforts. However they have not been sustainable, due to lack of solidarity about needs being met among all races, classes etc. (A special dream I have is that there will be many yearly Million Man Marches across the planet fueled by commitments to <u>sustainable</u> loving, healing action).

We all want to feel loved and secure, appreciated, and that we are making a significant contribution.

With individual and collaborative acknowledgment and ownership of the problems (Part I of this book), and commitment to coming together for healing, Loving action (Part II of this book),....we can, for sure.....mend our rifts, so that everyone feels worthy and valued. There is certainly enough wealth to go around...so that no child, woman, or man need go without food, clothing, shelter, healthcare, education....and that the animals and earth can be protected.

Continually looking deep inside myself to unravel and heal the layers of racism, sexism, classism, homophobia, xenophobia and other divisiveness that separate us, is one of the great gifts of my life.

I hope that you decide to join together in this incredible moment in time...the wealthy men at the top, hearts and hands with all the rest of us.....we can all have a life filled with the beauty and sweetness we each so deserve. The time for courageous choice …...is NOW.

With Love and Respect,
Roberto

Vulnerability and Strength.

First off..... as I am taking a deep breath.......I need to identify myself as a hetero white male ("pseudo" white male, to be more correct, as I have always related more to being "ethnic" rather than white, and "primarily" hetero, since I relate more to being Two Spirit... more on both later). Yet given all the privileges that I have with this lighter color skin, and not having to fear being persecuted for my gender identity, I feel it's necessary to identify as hetero white male.

And I feel the need to acknowledge .....that it's a whole lot easier for me to be talking about being "vulnerable", when I don't have to feel subjected to the overt, personally directed threat of racism every day of my life.

I have had some men of color respond negatively to my talk of vulnerability. "Yeah, easy for you to say, you don't have to be worried about being profiled". Or, "There is no way I'd teach my son to be "vulnerable", in a society that already treats us like second class citizens".

Ok....so that's what I initially wanted to say.

And... next...... let's ..please.. be clear. Vulnerability, as we are discussing it here....is **..not ..**about weakness or being taking advantage of, running from bullies, allowing ourselves or those we care about...to be intimidated, hurt,

<u>etc.....especially from those using their racism, sexism, homophobia, transphobia, classism, xenophobia, or any other excuses to dis-respect others.</u>

Some stories that follow, to further help clarify.

Story #1.

One of my respected and trusted racial justice allies and friend spoke out in a meeting about his anger at how protective he felt for the safety of his son, especially if he might be stopped by police. This is a fear he felt was very common to most Black parents. As a Black man of slave ancestors... and someone who had personally been subjected to racial profiling.....he was understandably very concerned and protective of his son especially regarding any possible interactions with police. So he continually took extra care in supporting his son, and reminding him to treat all others with respect......<u>to stand up for himself... intercede when others were being bullied,....not take crap from anyone</u>.....yet to be very very respectful with any interactions that might occur, with those in authority...especially police. This included specifics like, keeping hands on the wheel in a traffic stop, "yes officer, no officer", complying with requests respectfully, etc....taking extra care to not jeopardize his own safety.

Story #2.

I was offering a training for staff of a local outpatient alcohol and other drug treatment center on how male issues were related to alcohol and other drug use/abuse/addiction. One of the staff, a single parent originally from Puerto Rico, spoke of how, while she agreed with some of what I was saying about the importance of teaching and modeling for her sons healthy and respectful emotional expression....she was extra concerned about her sons being hurt and taken advantage of if they were perceived as "vulnerable" or weak. I expressed support and understanding for her concerns, and added...."this is about teaching them to be strong, able to take care of themselves, and to stick up for others, when appropriate....... and also to be confident enough in themselves,and their values that they are willing to be humble, sensitive, kind, gentle...etc...the highest qualities that would allow them to be their fullest selves. Strong and Gentle. Brave and Kind.

Story #3.

Last year, when food shopping, a man noticed the Veterans for Peace hat I was wearing, and said, "Veterans for Peace, isn't that an oxymoron?" We both smiled, and sat down on the window ledge to chat. I liked him right away. He is a wonderful, caring father originally from

Haiti, who shared with me his challenges, in raising two <u>strong, honest, brave</u>, young Black adolescent males..... yet with the highest priority being.....them not losing their inner "<u>soft hearts</u>" in reaction to a world that was often a "<u>hard</u>" place. Again, a seeming dilemma. ... and / or.. an opportunity to model strength.. and... gentleness as complementary....and on the road to real freedom and empowerment.

Story #4.

Yesterday I stopped and asked for directions from this very pleasant Police officer. We got to talking....he was helping to provide security for a gathering at a local Jewish Temple. We spoke about the need for security at the temple, in schools etc. Then I mentioned my book, and he told me his parents came over from China... language, and assimilation was a challenge, and he was raising 3 young boys. He said he'd really have to think about this vulnerability thing. He got the importance of them having "soft hearts", however making certain they would be strong enough to defend themselves when necessary would be essential.

(The next story is the reason I decided to add this Vulnerability and Strength section, after the book had already been out in it's first printing, because it was clear to me how important it is. With Great Thanks to You My

Dear Brother J., for your vulnerable and strong truth telling.)

Story #5.

I recently gave a copy of this book to a friend, a very wise, elder Black man, who I like and respect very much. A week or so later, I asked him what he thought about the theme of the book. He said...."This vulnerability thing.....I have a friend who runs a fruit stand, and he was threatened by some white guys who wanted to steal some of his stuff. You saying he should be vulnerable and let them get away with that?" <u>I felt myself getting a little defensive,</u> when attempting to explain the difference between standing up for ourselves when being bullied, and being emotionally vulnerable to build trust in relationships (of course in both instances there may be risk of getting hurt... primarily...physically in one, emotionally in the other). Later that week I acknowledged my initial defensive reaction to him. The irony of me writing this book about being vulnerable, and yet being defensive to his important point, was not lost on us. We nodded and smiled at each other.

Story #6

My same good friend from the previous story told me today..."I got another one for you. Just heard a young women in Central Park got attacked by a guy trying to rape her. She kicked his butt good...left him hurting. She came away with some bruised knuckles.. brave and tough". We both looked at each other and nodded, again.

Vulnerability...................Strength....................
Vulnerability...................Strength.........

Ahhh....Continually Choosing to be brave enough..... loving enough...... to lean in to the exploration of our common hu-man-ity. This stuff can be challenging........ uncomfortable.........messy......and yet.........soo rewarding. (With great love and appreciation to my Dear BFAM Johnny for his support in fleshing out this Vulnerability and Strength section)

Much more on Vulnerability and Strength to follow.

Mitakuye Oyasin / Lakota for All My Relatives, All My Relations We Are All One

A poem to all my relatives
From A white man on the REZ

I feel like
I've never fit in
And <u>that's</u> my only sin.

And to live my truth
need to Speak my truth
That we're each unique..
and each KIN.

My wounds teach me
where I came from,
Scars remind me
to never forget-
So many lessons to learn,
from folks I still haven't met

A song of surviving?
NO...I choose thriving
truth, not lying,
nor continual denying.

They say we're post racial
and let's move past-

I can't, and won't
it's in my bones, my cells, my molecules, my DNA
So healing don't come fast.

How many lifetimes
does it take to heal?
From being the only ones
forced to come here....
and being treated as subhuman, savages, slaves, animals, property.....
How many lifetimes to heal?

I don't want to forget that truth,
the values, that founded our country are still alive and well
elitism, power, wealth, control,
entitlement, competition to be #1 at all costs,
white heterosexual male privilege
all values that can create living hell.

I have a mental "dis-order",
that needs to be controlled?
Or is the system what's disordered
and I just don't fit the mold...
of the values that are cold.

If I say I don't see Color,
I'm really saying I'm blind,
to the truth of having a dark skin
and how folks can be so unkind.

If I cut my hair
speak, dress, walk and talk, WHITE,

then, maybe the American Dream will include me,
am I dreaming? Wake UP!!,
they really want me disappearing from sight.

For we can't sing and dance together,
and we can't be truly free,
until we acknowledge the truth,
of what I can't always see,

That we are different,
and we are the same,
and the tragedies
are not about blame.

And the Tyler Clementes, Travon Martins, and Michael Browns of our country,
will continue to be murdered like dogs,
Until we stop pretending their human,
and admit they're still seen as "hogs".
(from "A Lesson before Dying" by Ernest Gaines, a young Black man in the 40's South, wrongly convicted

of killing a white woman, after being called a hog by a
prosecuting attorney, said before he was executed, he just
wanted them to know, he wasn't an animal, he's a man.)

Unless we're willing to remember,
what was done to the Indians and slaves,
then we might as well dig us a Hole,
and go back to living in caves.

It's too uncomfortable, you say,
let move on to a brand new day,
not nearly so uncomfortable, I say
as spending a life playing straight,
when I'm gay.

So simply being alive,
Do I deserve a piece of the pie?
Or do I need to prove my worth,
by continuing to live a lie?

Hmmmm.

So what's the answer,
I ask,
It's clear,
No easy task,.

And yet, it's really quite simple,
the wise ones have always taught,

That All life is sacred,
and worth is not earned or bought.

That each of us is born worthy,
and All life is precious and unique.
Yet each of us are also connected
and deep, deep inside is All we seek.

For the greatest power is truly LOVE,
yet rarely is that what's spoken,
in our homes, churches and schools,
in our counseling rooms even a token.
What are we so afraid of,
to be all we truly can be?
By teaching each other bout love
the essence of you and me.

And our boys to men
in each culture,
will Then start, learning and grow,
to be strong and gentle and honest,
and be more than we'll ever know.

So please, My Dear Friends,
please, please look deep inside,
for we each are precious and sacred,
and there is nothing to hide.

For the real adventures
aren't out there.
The real adventures
are home......in here (in our hearts).

So let us embark on this day,
with hope, courage and joy,
And celebrate the backgrounds
we come from,
As we tender the man from the boy.

Mitakuye Oyasin.... Namaste.....Molte Grazie.
(This poem was part of a keynote for a multicultural counseling conference entitled "Boys to Men" in Multicultural Counseling: Envisioning Change through a Male Lens", November 2014.)

Letter 2.

Brothers,

This book is my offering in support of each of us, coming together, to finally free ourselves from the "control" of the incestuous dis-ease of machoism....and all the other "isms" it fosters....racism, sexism, classism, xenophobia, homophobia, materialism, wars, killing animals, destroying Mother Earth.

The macho value system is another term for wealthy white hetero male supremacy cultural values, which started this country and still runs it....and therefore strongly impacts each of us. Some of those core values are power, wealth, control, competition to be number one, at all costs, no matter who we need to walk over to get there, materialism / success, elitism, entitlement. These values tend to give us a false sense of "security"/ "safety", when fear of losing our tenuous hold on them threatens our health and well-being, and leads to lives filled with "stress" related dis-ease. The "quick fixes" we have been sold, are used as substitutes from learning how to truly care for and comfort ourselves. Thus we move further and further from our truest, highest selves, and become more and more dependent on things outside of ourselves.

Honestly and humbly facing the powerful influences these macho values have in our lives takes great courage. I don't, for a second, suggest this is easy work. To the contrary, this is not for the meek of heart....and is often excruciatingly painful....yet can be so incredibly rewarding....if we but make the initial commitment, and then stick to it, in the midst of seemingly impregnable barriers.

If we choose to fight to defend and maintain status quo... as many of us have always done......clinging to our hurtful macho wealthy white hetero male cultural values......then we continue to stay alienated from ourselves....our truest and most beautiful selves, and all others.

Devoting our lives to this work helps us to finally return home again....where we belong.....safe, and loved ...in the womb of Mother Earth........congruent.....at peace.....in Love....with ourselves....and All Life.

It is my prayer and hope that each of you who are reading this, find this small book to be helpful in your own healing journey.

May Love... Courage, Humility, and Integrity guide us on our way.

Roberto

Fear of Tears.... or......Honoring Our Tears.....A Truly Radical Path to Healing

I know this is a dicey issue for most of us. So please bare with me as we gently traverse what I believe to be one of our greatest paths to healing.

Hmmm...why is this such an uncomfortable topic? What are we so afraid of?.......

.......we'll be called a baby...wimp.....weak....a girl ... and on and on. We'll just be a deep puddle of tears.... and never get out of it to be the strong person we need to be.....

........or we'll just be giving in to our weakness, and never do what we need to do to protect ourselves from all the threats <u>out there</u>......

(or is it really ....<u>in here</u>.....in us).

Or you might say...."I'm not afraid of my tears....I just don't have any." And yeah..I get that. And........that's sad... and maybe more than a little scary. We men have been taught to not cry.

"Stop crying, or I'll give you something to really cry about". While it may sound contradictory.....I have gratefully learned...that a major part of happy lives, and emotional health....can be our connection to, and celebration of, our tears (which, of course, can also be tears of joy).

When we wall off from our tears....we wall off from a precious part of our hearts.....we learn to hide behind a suit of armor...to protect ourselves from being "vulnerable"...to all the other "threats in our lives".....we cut ourselves off <u>from ourselves,</u> and deprive ourselves of one of the most important, powerful, and <u>essential</u> aspects or our hu<u>man</u>ity ...of <u>Being</u> an emotionally healthy, balanced...both strong and gentle....<u>real</u> man (real men Do cry).... . (In his powerful video, 'The Wisdom of Trauma', Gabor Mate teaches us, the greatest trauma is the trauma of being alienated from ourselves).

The importance of tears as a pathway to healing cannot be overlooked. We don't just get over childhood trauma, by just wishing it away. The wounds stay buried, deep inside our minds, our emotions, our bodies....each cell, each molecule, each organ....festering....hurting us more and more......until we finally start tending to them. "Just get over it....be a man....that was then...move on," doesn't work. Until we take the necessary time to do the healing work, we won't heal. And the longer the wounds go unattended.....the longer it takes to heal. Our tears can be such a great, soothing balm for our pain.

Often times we don't even allow ourselves to cry at funerals...Be A Man....Be A Man......that is totally insane. Crying is how we honor our feelings, and how we grieve and heal our great losses. Yet we are often taught to .."just be strong".......and forced to "survive" without this amazing, life affirming...... soothing.... comforting...... nurturing......cleansing....... freeing.....<u>healing</u>.....gift......... of crying........honoring and then releasing deep sadness and grief.....which is about as natural and <u>free</u> as we can get.... (along with, of course, crying over great joy.. another part of being truly alive and "thriving").

And we wonder why so many of us men are emotional basket cases. This is usually due to our tenuous hold on our emotions.... our lack of healthy role models and teaching...and the unhealed pain from our childhoods. (What's really scary, is we often have "adult" leaders who are really unhealed little boys, running the show)......... So we strive to be stoic robots....(yet are allowed to bully with anger...or even act out in rage)....and we wonder why we have racism....and violence against women children, animals, the planet.... and ....war! Speaking of war... the unexpressed trauma, guilt, shame, sadness, fear that many of us vets hold inside from war atrocities we can't share about, for fear of appearing weak, are often the cause of extreme violence, suicidality, addictions and other extreme mental health issues. Crying, leading to

Men and Racism: The Healing Path | 21

sharing openly, would have been a great outlet and healer for a lot of us vets. And the rage, which lives inside us, and causes so much of the horrific atrocities in this world, is most often the result of our untended to tears.

When we tend to our natural inner need to cry, we can feel more at peace with ourselves, and our bodies and minds feel healthier....because we are listening to our inner voice. Many health care practitioners recommend crying to keep our bodies well tuned, and free from disease...emotional, mental, physical. After some great and tender cries....we can feel ….soothing relief.....cleared out....maybe even ready to start the process of eventually letting go of old losses, wounds, and fears that no longer serve us..... We can learn to start making healthy loving decisions... followed by healthy loving action....perhaps by asking ourselves...."What would love do now?" as our guide. This doesn't mean we won't still grieve some pain and losses, maybe for the rest of our lives. It does, however, mean we can also decide to savor the sweet memories and feelings, heal and appreciate who we are, and what we have now, and in the future.

My fear of crying governed half of my life. I know that my absolute terror at being found out to be less than.... unworthy.....not smart enough...not good enough.....a liar.... cheat...fraud...weak...a coward.....was the fear that drove so many of my decisions and actions for at least half of my life. Finally in my mid forties, after being suicidal, I got into some serious treatment and recovery and decided that if I was going to choose to live...... I would need to make a life long <u>commitment</u> to learn about real love, and how to really respect, honor and love myself and others, by living in complete integrity,and by facing all of my fears....especially of being vulnerable. So I started doing some deep grief work, allowing myself to cry, which I hadn't since early childhood (to be a man), along with deep rage work (i.e., yelling at the top of my lungs, beating cushions etc, <u>however being fully committed to not hurting myself or anyone else with it</u> (didn't need more self-inflicted hurt)). As part of my self care routines, I'd set times aside, when I wouldn't be interrupted, and would have some time to rest after (sometimes I would actually feel energized afterwards). Along with meditation, prayer, and journaling, breathing and repeating positive supportive affirmations.... little by little I finally started feeling free of the deep dark block of granite pain in my gut, and started actually liking myself.....and not so afraid of being vulnerable ....actually seeing it as a path to being the man I truly wanted to be. Now, when I see something, or read, or hear a song, or am out it nature and some sensation, or sad memory or thought comes up, and I allow myself to cry....I feel so grateful....so Gratefull. (Please listen to and see the beautiful and moving song Grateful....by googling "<u>you</u>

Men and Racism: The Healing Path | 23

tube grateful tony moss lyric video" (and click on the little box in the corner for full screen view).

As a long time trauma therapist, these days I only work with male trauma clients. And learning how to be gentle and kind...tender-hearted with ourselves, continues to be one of the greatest challenges for most of my clients. Yet, when we are brave enough, to gradually let go of the suit of armor, we start to realize that most of the things we've been looking for outside of ourselves......actually lie inside us.

Our unattended to scared little boy will continue to hide inside us, in deep pain.. forever affecting our "adult" lives' feelings and actions.......until we finally begin to listen to and start comforting and healing that most vulnerable part of us. <u>Choosing to become the source of soothing safety, acceptance, comfort and healing love for ourselves, frees us to finally truly be all that we can be</u>.

Years ago there was an amazing love song I heard....(wish I could remember it and find it)....it basically said....the greatest gift I can give you....is the gift of my tears. Makes me re-member...some of the cherished moments when I have felt closest to another human being were when we held each other and shared our tears. And yet, unfortunately, those moments have been all too few. For me, this is real emotional intimacy... where real trust and safety and love can flourish. As long as, of course, it is part of an ever growing commitment to core ingredients of love......like gentleness, integrity, courage, humility, dependability, kindness, compassion...... Some wise one said...."Each tear is a another drop of healing....on this life long adventure of healing".

This book on 'Men and Racism....Fear of Being Vulnerable'.....can be most complemented by a companion guide on the essence of healthy crying...which might just be my next one. Or maybe one of you will decide to write it. In the meantime, for more inspiration about the gift of our tears, please see the work of Henri Nouwin, 'Men and Grief', Richard Rohr, 'The Gift of Tears', Dr. Kate Truitt, 'Keep Breathing', and my Dear Friend Dr. Amanda Aminata Kemp with Dr. Sina Smith ('Why is Crying Good'/'Why is Fall the perfect time to grieve', on you tube video, which reminded me of how important crying is for our health...duh... I wouldn't still be here without this amazing gift of my tears)

Macho Man..from boyhood to man.

The refusal to succumb
renders us all numb
from **feeling** who we are
alienated from ourselves... so far.
(the first line was originally, "The root of all dumb",
because, while I do feel macho is dumb, because it
hurts us all, I didn't want to make any of my brothers
feel insulted or defensive, or disrespect anyone's culture
traditions, including my own.. so I changed it to The
refusal to succumb, meaning to give in/ be vulnerable.......
however......).

From the time we're a child
we're taught to be wild
or intellectual, number one
is what we're molded, to be fun

To prove our worth, the stronger,
a crybaby no longer.
Tender feeling is to be weak
walled off from gentle
is what we seek.

To always be at the top
Otherwise we're a flop.
Competing with each other,

instead of each our brother.

Sets us up to fight and pretend
we're in control until the end...
over women, animals, the land
no clear picture of healthy man.

All others perceived as a threat
don't let them see us sweat.
Racism, sexism, homophobia,
refugees, cleansing utopia.

What a treacherous, treacherous game,
It's important to finally name.
For we'll never be truly alive
staying stuck in this race to survive.

And if we really want to be free,
we've got to take off the blinders to see
that we've been hoodwinked all along
and being macho is all wrong.

It's a front, a lie, a sham,
Being gentle and kind's who I am
Being brave and honest and humble
Admitting I'm wrong when I stumble.

Committed to equity for all
prevents me from the "great fall"

from being all I can be
loving human...like you and me.

Being nurturing, considerate and kind
we're finally out of the bind
Then we wake up
and shout with joy

As we cull the man
from the boy.
And we know that fully living
is to share, and to be giving.

Each one, our sister, our brother
love inside for our self and each other,
giving love to ourselves and each other
giving love to ourselves and each other.

12/31/19

Men and Racism
The Compassionate Path To Healing

While we men have made so many wonderful and truly amazing contributions to this world..........unfortunately we have also been responsible for the vast majority of the horrific atrocities.....the extreme violence, rape, wars, desecration of the animals and the land.......sexism, racism, homophobia, classism, xenophobia. And while women are capable of the same....for the most part they are emulating the example set by us....in an effort to achieve an "equitable" piece of the pie.

So there you have it. This is what I need to say....up front....so you have no doubt about who I am and how I feel about all of this.

Racism can be described as a system of disadvantage based on race-
thus the term white supremacy -which founded and still runs our country. As far as I can tell, at least in this country, it's been living by the wealthy, white, hetero male supremacy cultural value system that has caused and continues to sustain our racism. (While I don't have a lot of money, I consider myself "wealthy", given that most people around the world live in such dire poverty). And, I do believe, that theoretically, depending on our, familial

and cultural upbringings, we all can see folks of different races as a threat, leading to discrimination, biases, and horrific treatment.

Please know, and continue to be reminded…..I bare none of us ill will….guilt or shame…. as this isn't helpful. I've done enough of that to myself.

I'm all about us healing the darkness inside, so we can have the beautiful life we all deserve. I say "we"/ "I"..; **because I hold myself accountable, and claim my part in all of the aforementioned horrific atrocities… because I know that the evils of sexism, racism, homophobia, classism, xenophobia live deep inside each of us, especially growing up in the white supremacy value system in this country.** How could they not. From the beginning of time…we men have struggled with feeling like we had to prove our worth ….by accomplishing things, and accumulating material things, and often seeing other men as threats rather than allies.

The answer, as I see it…..it really quite simple…..yet exceedingly "radical"……..It's about reclaiming the essence of each of us……that each of us….and All Life…..is **SACRED.**

Not talking here about religion...as that has also been a perpetrator of both good and evil....so that often turns off many of us. Even though the irony is that I do believe that the essence of most religions also teaches that all life is sacred. However, that core message, all too often, gets lost in the translation....and diluted by men who want power, wealth and control. So the religions that teach about the Sacred as a Spiritual concept...the Loving energy of the universe....that we are All one, all connected, "All God's Children", would be great supporters of efforts for equity and justice for all.

<u>When I talk of the Sacred...I refer to ......All Life Being.....important, valuable, worthwhile, precious, cherished, unique, worthy of Love, dignity, respect, awe and wonder. That All Life is Sacred....and I'm no better no worse than that insect, that plant, that tree, that animal, that woman, that child....and that other man. ALL LIFE IS SACRED!</u>

Just imagine how different we would be...how different our country...our world would be ...if that was the core principle we taught every child......and that is the core principle each of us chooses to live by....and that is the core principle....all our leaders.....and all institutions lived by. And that as a reflection of this core belief in the Sacred....that most powerful force in the universe …..Love.... is taught as the root foundation of All The Sacred. More about the Sacred and Love to come.

You won't necessarily agree with everything in this book. So, as we say in 12 step rooms...please feel free to "take what you want, and leave the rest".

~

P.S. My beloved friend, mentor and teacher of Traditional Sicangu Lakota Spiritual Teachings, Dr. Tom "Rags" Balistrieri, stopped me at the door of one of my first sweat lodges, and said..."Roberto, you're such a good man...but you're so hard on yourself.....you've got to learn how to be compassionate with yourself."
Here's to each of us learning, and committing to this all important lesson...for without it....not sure about the possibility of real healing...for ourselves or anyone else.

white supremacy culture

"Why insist on using the term white supremacy when it is so volatile and gets folks so defensive?"
Dr. Robin DiAngelo explains it best in her seminal article entitled, "No I won't stop saying White Supremacy" (The Good Men Project, August 2017). She teaches that using the term white supremacy puts the onus where it belongs, squarely on the shoulders of white folks. <u>This is not about guilt tripping.</u> It's about taking honest, accountability and responsibility. We are the ones who have benefited from this system the most. And why White hetero male supremacy? Because the vast majority of our institutions are still run by White hetero males, and Women still only make 70 cents on the dollar compared to what men make (not to mention what folks of color make). There is nothing wrong with being wealthy, White, hetero male..it's what is done with that power, to insist on equity for all. It's my responsibility and it's our collective responsibility. And if those in power don't hold themselves accountable, and/or are not held accountable by others for acknowledging and then changing the inequities, which have existed from day one, things are never going to really change, other than with quick fixes. And by the way...poor and middle class White men are also getting short shrift.

(And teaching and modeling for our boys and young men...healthy honoring and managing of feelings, and kind and considerate behavior, is a significant lack.... which contributes to the above, and is addressed throughout, and especially in the sections on Fear and Vulnerability, and the All Life Is Sacred and Core Curriculum sections).

white supremacy culture

white supremacy culture, the <u>name</u> of the game
started our country, puttin us to shame
God's anointed ones, laid the first claim
doctrine of discovery, manifest destiny
for riches and for fame.

Justified the genocide
of the original Native ones
enslavement of African people,
chains, whips, guns.

The system had core values,
power, wealth, control
competition for #1,
no matter **what** the toll.

But that was then you say,
surely things have changed.
Things may **look** a little different
but only re-arranged.

For the wealthy 1 percent
still firmly at the top
and the precious young Black man
still **afraid** of being killed by a cop.

The system so deeply embedded
in our culture, from the very start
and now it's up to us
to cure this from the heart.

For reasoning is not enough
white arguments so clever.
It's up to you and me
to do this work forever.

To be honest and be humble
to leave no stone unturned
with courage and compassion
equity for which we've yearned

So thanks for being here
each one, your important choice
this is the moment in time
to heed our inner voice. Mitakuye oyasin
"Until the killing of black men, black mothers' sons, becomes as important to the rest of the country as the killing of a white mother's son, we who believe in freedom cannot rest." Ella Baker

On Being Vulnerable

A few words about this incredible concept.....men actually Choosing to Be Vulnerable....what a freaking Concept!!!!

Vulnerable.....what does it mean?.....and why choose it?

"What are you Crazy?"/////I can hear many of you saying....."Why would I Choose to be vulnerable?"

In this beloved crazy mixed up world of ours....that is usually one of the last things we men would consider aiming for. Maybe after we've been divorced a few times, and now in marriage counseling for our 3rd shot at a lasting relationship. And our partner says...."I want to hear your feelings, I want you to share your secrets.....Let me in, I'm tired of this macho bs." So the softer, gentler side of us, actually becomes desirable....and we are clueless....or if not clueless......petrified of letting down our "guard", which we have perfected to judiciously protect us for most of our lives. "Vulnerable you say".....I'll get laughed at, ignored, hurt, put down, rejected, walked over...and on and on. Yep...that sure is a possibility.

Letting our "guard" down...taking off the steel jock strap, opening our heart, sharing our intimate feelings, taking the risk to be let down, misunderstood, embarrassed, and worse of all rejected.

Yep that's sure some possibilities. That's why **it takes so much courage to be vulnerable**. "So wait", I hear some of us saying...."being brave to be vulnerable!?"... may seem like an oxymoron.......because vulnerable has gotten the bad rep as being for weaklings, who will be walked over. So, yes, I know it may seem counter intuitive.....big time.

Which is part of why it can seem so challenging.

And …..as some of you wise ones have found out... by choosing to be brave... the amazing gift of love...... is really only possible with this amazing choice ….to be vulnerable. And the truth I've discovered is that by choosing to really love and earn my own trust and respect by being honest and dependable and accountable etc., all the necessary ingredients for love…….then I finally truly embrace all of me…..no more having to hide or protect secrets etc…..**which opens the door for true intimacy with another.** I will only be brave enough to live this way….if I have consistently demonstrated over time that I can depend on my commitment to myself…..that I know I will always be there for me, no matter what happens. That is really all I can know for sure. All the rest is out of my hands. (Of course, if you have strong spiritual based teachings of love, that will also provide great support in healing the great divides that plague our planet).

Back to racism. Since racism …and all the other 'isms', as I see it…..come from fear, defensiveness, mistrust, disconnection etc., …when we are finally wise, brave and loving enough to open ourselves to others, because we no longer feel the need to keep the walls up….then will racism cease to hurt us and all others. We're all very imperfect humans...that is what bonds us....what helps us have empathy for ourselves and each other. If we start with having empathy, kindness, forgiveness for our self.... then, can we do so for our brothers.....and sisters....and the animals, earth....all life.

Vulnerability

**Vulnerability,
the antidote
to wealthy,
white,
hetero
male
supremacy**

Since the time
of our ancestors,
the cave "men",
we males learned
to view everything
with **suspicion**,
hatred
and **fear**...
keeping the softness
of love
buried
deep inside,
protected
in suits of armor
and wealth.

We've all
been duped...
to be alienated...
from ourselves...
and each other.

To see ourselves
and each other
as **threats...**

instead of
as supportive,
compassionate,
nurturing,

beings
of light
and love,

Beings
who are
trust worthy.

**This Original Fear
of all others**
is the foundation
of racism,
and all the other
"isms",

and intentionally
prohibited
 "We The People"
from ever
truly meaning
 "All the People"....
never really
 intended
to apply to
women,
children,
sexual orientations,
refugees,
races,
ethnicities,
abilities,
classes,
animals,
the land.

The good news!
We now have
 the opportunity
 to truly "evolve",
by embracing
our true wisdom,
and strength,

to be gentle,
and kind,
and brave enough,
 to be openhearted,
and vulnerable..
to feel deep love.

This willingness..
to risk..
being hurt,
is the essence
of real growth,
real love,
real men.......healthy humans.

A Synthesis and Re-Imagining

This book, the "last" in the series, is largely a synthesis / re-imagining of much of my first two books, which reflect an underlying theme ....namely ...the powerful responsibility of us men in contributing to the creation and maintaining of racism and all the other related hurtful oppression of sexism, classism, xenophobia, homophobia. However, in my efforts to "not offend" or contribute to defensiveness...(even in my second "uncensored" book), I soft peddled the **unhealthy male values theme**. I didn't clearly enough name it, for what I believe it is, the core cause of all hurtful bullying, (be it intellectual, physical, emotional, racial, sexist etc.). So I have added other pieces, which I believe, further help our efforts to address these challenges.

I offer this "guide book", to hold myself accountable, to offer support to you...and to further clarify the importance of committing to this work. I so passionately believe that this is where the hope lies, if we are to truly heal ourselves and our world.

You might say to yourself, "Who does this guy think he is..that he has "the" answer which no one else has come up with?". My response to that is....I think many have suggested and alluded to this theme....I just think we may have been concerned about offending, or not including

Men and Racism: The Healing Path

women and all other oppressed groups..etc..etc.... So **to be clear...I am not leaving out anyone....as we are all needed ....and... we all are responsible.**

However, I believe that the few wealthy, white, hetero males at the top...the 1%ers....still have, and will continue to have the power wealth control....and are master manipulators in placating the masses every time we rise up to protest and demand equity and justice. So compromises are made, and improvements enacted.....the masses settle down (often through shear exhaustion and relief, and satisfaction that finally some change is happening). And for a while things seem better....and then, before you know it....bam..... the same shit is happening...although at first glance it may seem a little different....the oppressed are still being oppressed,...and we all are getting screwed by "the man." "Well yeah, you might, begrudgingly say, "I see that, ...but.....it's still better here, than anywhere else...so stop complaining and be grateful and patriotic, and shut up so you don't threaten what we have!"..

<u>I continue to believe, that significant, **sustainable** changes which ensure real equity and justice for ALL, can only happen if the few wealthy, white hetero males at the top are finally held accountable, and ideally, choose to acknowledge their lead role in this mess since</u>

the beginning. And after some serious soul searching, (as in 12 step programs, a fearless and searching moral inventory), after considering sustainable healing strategies, vetted by those so hurt, then perhaps offer a sincere, heart felt amends, and proposals for sustainable change, that will be continuously checked by community groups using **values check list inventories** (please see rudimentary, sample inventories for citizens, leaders, institutions etc., in the All Life Is Sacred proposal section).

The following was used as part of a keynote presentation on "Men and Multicultural Counseling". The belief is that, as we support our boys and men, and all of us, to embrace and live by our most healthy human values, racism will eventually die out.

Values

It's About Values
And Taking A Stand

This being human,
Always a choice.
When to take a stand,
and lend our voice.

Knowing our values
Being firm and clear,
Always a choice,
between love and fear.

So let's see now....

"Unhealthy",
    traditional,
white,
    heterosexual,

<u>**masculine,**</u>
    <u>**supremacy**</u>
  <u>**values;**</u>

competition
    elitism
entitlement
    superiority
power
    wealth
success
    control

Or,

<u>**"Healthy",**</u>
    <u>**traditional,**</u>
<u>**feminine,**</u>
    <u>**(and "healthy",**</u>
<u>**male,**</u>
    <u>**values:**</u>

cooperation
    sharing
generosity
    consideration

nurturance
    kindness
support
    encouragement

And,

"<u>Healthy</u>",
    <u>traditional</u>
<u>human</u>,
    <u>values</u>:

integrity
    dependability
courage
    humility
gentleness
    strength
service
    respect
equity
    patience
compassion.

**Alright now...
hope that helps you some
always important to figure
where we're coming from.**

Overview

Racism Men
and the
Fear Of Being Vulnerable

An Initial Guide For Healing Ourselves and Our World

If you care about the racism, sexism, homophobia, classism, xenophobia, the animals, the earth, and all related problems of our world....and want to do something.....really...do something.....what follows is an initial healing guide for us all.

First we will briefly address, what I believe to be the most significant, contributing factors to all the "isms". And then we will look at the antidotes.

Interwoven throughout the rest of the book, are streams of consciousness, poems, short essays, proposals and other pieces that you will hopefully find supportive, invigorating, and helpful on your own healing path. They are the outgrowth of my own healing journey, and the incredible lessons I have learned from knowing and working with so many incredible women and men as a trauma / racial trauma therapist.

So here are some of the most powerful inter-related contributing forces to some of our greatest problems:

1. TRAUMA – emotional physical, reaction to severe stress and threat to our well-being, can be one time, repeated and/or inter-generationally.

2. DISCONNECTION – feeling separate from, isolated, from ourselves and whomever and whatever helps us feel at one with ourselves and others.

3. FEAR / ANXIETY / STRESS – feelings of worry, confusion, doubt, dread.

4. PAIN / HURT / SADNESS – the emotional wounds living inside us, and resulting scars that may live on for the rest of our lives

5. ANGER / RAGE – another normal reaction to the pain of the above, deep pain is usually the core of anger, and rage the result of unaddressed anger and injustice.

6. INSECURITY – not feeling stable, calm and reassured, inside ourselves, and in the presence of others.

7. MISTRUST - suspicion of integrity, dependability of ourselves or others.

8. LACK OF SAFETY – fears about our emotional or physical well-being alone or with others

9. LACK OF APPRECIATION – not feeling good enough, or well thought of.

10. LACK OF AFFECTION AND SAFE TOUCH – not receiving necessary physical and/ or emotionally soothing connection

And here are some of the simple, exciting, and extremely challenging antidotes to the aforementioned contributing factors:

1. HEALING TRAUMA – acknowledging the truth, reassuring self, and seeking support for normalizing reactions, and addressing the following:

2. RECONNECTION – seeking ways to remember who we truly are, and feel grounded and

reassured in that truth, in relation to ourselves and others.

3. COURAGE – choosing to face our fear with loving action

4. COMFORT – reassuring ourselves and taking in support from others

5. RELIEF – from emoting and receiving acknowledgment from self and others

6. SECURITY – feeling at peace in presence of self and others

7. TRUST – feeling can rely on self and others

8. SAFETY – ensuring healthy emotional and physical boundaries

9. APPRECIATION – feeling worthwhile, affirmed

10. AFFECTION / SAFE TOUCH – welcomed, gentle emotional and physical connection

A very brief yet essential hint to the essence of all the above mentioned factors and antidotes which we will further dive into in the ensuing sections:

LEARNING HOW TO,

AND **COMMITTING** TO **100%**,

   ALL IN,

WITH EVERYTHING WE GOT.......

TO …..**LOVE**….
OURSELVES

AND

EACH OTHER

AND

ALL LIFE-COMMITTMENT

WITH ALL THE ESSENTIAL IN'GREDIENTS

OF **LOVE**...

WHICH WE HAVE NOT, FOR THE MOST PART, BEEN TAUGHT AND / OR HAD MODELED FOR US. THAT'S THE TICKET FOR ALL OF THE AVOVE PROBLEMS AND ANTIDOTES.

THIS IS NO EASY TASK. AND WILL BE CHALLENGING...TO SAY THE LEAST....AND

YET....AND YET.....EXTREMELY.....GRATIFYING....
LIFE GIVING....LIFE AFFIRMING........HEALING.....
FOR EACH OF US ....AND OUR PLANET......OUR
UNIVERSE.

WILL YOU BE HUMBLE ENOUGH, AND BRAVE
ENOUGH, AND COMPASSIONATE ENOUGH....TO
TAKE THIS ON?

THAT IS THE QUESTION.

THIS IS NOT FOR THE MEAK OF HEART.

TO THE CONTRARY.....FOR THE BRAVE HEART......

THE WARRIOR.....

THE ONES WHO ARE COURAGEOUS ENOUGH

TO COMMITT TO BEING 100% HONEST /
IMPECCABLE WITH OUR WORD

AND TO INCREDIBLE EMPOWERING LIVE
GIVING CHOICE OF **BEING** …...
V U L N E R A B L E.

WO....NOW THERE'S THE RUB.....HOPEFULLY A
GENTLE KIND SOOTHING RUB OF OUR HEARTS
AND OUR WHOLE BEING.

BECAUSE WITHOUT THAT **COMMITTMENT TO LOVING VULNERABILITY,**

IT SIMPLY ...AINT GONNA WORK!!!!. I DON'T NOW EVERYTHING...BY A LONG SHOT....BUT I AM CERTAIN OF THAT.

TO HEAL OURSELVES AND OUR WORLD **IS POSSIBLE**....I DO BELIEVE....

I KNOW IT....AND I DO BELIEVE....SO DO YOU.... DEEP IN YOUR HEART OF HEARTS......

BUT THERE SURE AS SHOOTIN AIN'T A CHANCE IN HELL / HEAVEN IT WILL HAPPEN....WITHOUT AN ALL OUT, ALL IN COMMITMENT TO LEARNING HOW TO LOVE OURSELVES AND EACH OTHER WITH LOVING VULENERBILITY.... WHICH IS THE GREATEST GIFT IN THE UNIVERSE.

AND THIS IS WHAT THE WISE ONES HAVE BEEN TRYING TO TELL US SINCE THE BEGINNING.

I SURE CANT CONTROL ALL THAT HAPPENS IN THE WORLD.....BUT THE ONE THING I CAN CONTROL....SHOULD I CHOOSE TO......IS THE ATTITUDE OF GRATITUDE I BRING TO EACH MOMENT...WITH HOW I TREAT MYSELF AND ALL

OTHERS WITH ALL THE BREATHS, MOMENTS, OPPORTUNITES I HAVE LEFT.

HERE IT IS. THIS MOMENT. THIS BREATH....THIS OPPORTUNITY TO CHOOSE.

~

AS MARGARET MEADE ONCE SAID...SOMETHING LIKE...

-NEVER DOUBT THAT A SMALL GROUP OF DEDICATED INDIVIDUALS CAN CHANGE THE WORLD....INDEED IT'S THE ONLY THING THAT EVER HAS.

Native

I embrace Being Born of the Sweet Mother Earth

I embrace Being Child   I embrace Being Adult

I embrace Being Female   I embrace Being Male

I embrace Being 75, at least in this life

I embrace Being Italian, Francesco Roberto Vincenzo Schiraldi, roots in Palo del Colle, and Marsico Nuovo (just wish my parents spoke Italian at home, not only when we went to my grandparents), and I honor my ancestors (and yours), and the history and traditions we lost ...in order to assimilate)

I embrace Being from Brooklyn and East Rockaway

I embrace Being working class.....and educated.

I embrace having these dark skin pigmentations- "beauty marks" (as my grandma called them)...all over my contrasting lighter skinned body.

I embrace Breathing.......And invite us all to Breathe.....

I embrace Being Lakota

I embrace Being African

I embrace Being Latino

I embrace Being Asian

I embrace Being 2 Spirit/Gay/Trans/Straight

I embrace Being able bodied and having disabilities

I embrace Being The Butterfly, The Buffalo, The Gorilla, The Lilac, The Mountain, The Ocean, The Fire, The Sky

I embrace Being Unique....I embrace Being Connected to the Heart of Each of You......
I embrace Being One, With All Life, and.. Breathing... Being Fully Committed to uncovering the white supremacy conditioning in me.. and in all of us. Breathing....Being Imperfect and willing to keep learning from my mistakes in doing this difficult work, with courage, integrity, gentleness.
I embrace Being Alive....... I embrace Being Me.. Here with You.
Pilamaya, Grazie
FRVS / Summer 02, revised 10/14,11/17

STORIES

OF

FEAR

THE SACRED

LOVE

AND VULNERABILITY.

"You been had, You been took, You been hoodwinked! Bamboozled! Led Astray!"

            Malcolm X

FEAR.........Back to the beginning.....Some views and stories about Fear..... the Sacred....and Love.

FEAR Intro.....we live in a …. "Culture of Fear" with ...violence, wars, economic stress, poverty, wars, climate change, insecurity and mistrust ....surviving..... rather than thriving ... with Fear Based Values Systems... get more, better than, more powerful then, protect what "we own".... "Ruling" our world. How did this happen?

Story 1. The Garden of Eden story. Many years ago....a dear friend shared a story she wrote about the Garden of Eden for a feminist lit course she was taking. Simply put, she flipped the three key roles. In "her" story.....Creator is a Woman...(Wo.Man ...as my beloved partner Eileen likes to refer to it. I use the present tense "likes to" ...as...while she passed five months ago......Eileen is very much alive inside me.....as I feel, think and write these words. I will share more of her soon). And Eve is a Powerful Woman who believes that Mother Love Creator very much wanted and wants us all to have the core knowledge that All Life is Sacred....and that Love which lives inside each of us...is by far the greatest power in the universe. However, Adam....felt very vulnerable, exposed, naked, afraid....was very afraid to have all this knowledge...because he knew the implications... the responsibilities necessary to actually Thrive in Sacred Love. So he enforced his physical strength on Eve.....

and guess what......men have been enforcing that physical "strength" (as well as "mental" and other "strengths'" ...ever since.....thus the state of our world. Again, please be reminded, this is not to discredit any of us, especially when we have stepped up, with loving courage to do the right thing, which has obviously happened repeatedly, otherwise we wouldn't still be here. However, this is about me, and you, and all of us, continually committing to being the most loving versions of ourselves now and with each breath, for the rest of our lives.

FEAR - Fear can be seen as a normal response to a one time and/or persistent stressor. However it can even lead to immobilizing or disabling us, if we don't address it. And.. it can also be viewed as a wonderful signal...a wake up call, if you will, that loving action is needed. Loving action is always the great antidote for fear. Loving action will provide reassurance, comfort, relief. All by responding with the simple, yet so powerful question........ ....**"What would love do now.?...."What would be the most loving thing I can do for myself in this moment?" (from Neal Donald Walsch, in his 'Conversations With God' book series). It will often be a little voice, from deep inside.** So we need to practice, gently quieting our noisy often distracted minds, with calming, soothing breaths, to really hear this freeing, life giving, life affirming...loving response ("Mindfulness

Meditation," Jon Cabat Zinn), is a an example of a simple, wonderful practice that helps us to objectively observe and acknowledge, thoughts, feelings, sensations, sounds, and gently, return "home" again...to breath. It is a regular part of my daily, morning, evening, and whenever I need it, self care practice).

F.E.A.R....FALSE EVIDENCE APPEARING REAL. That's the acronym I learned eons ago from from two wise, strong women who became my trauma therapists for my own initial healing work.

SO WHAT IS THIS GREAT LIE?....THE FALSE EVIDENCE APPEARING REAL LIE?......

SO MANY LIES WITHIN THE LIE.....i.e., lie # 1 we are born lacking....and lie #2...our worth needs to be demonstrated, by lie # 3 accomplishing and lie # 4... accumulating.. and on and on....and it's never quite good enough, so we have to keep on doing more and more and more ...in a never ending battle to prove ourselves worthy enough.... smart enough, strong enough, wealthy enough, powerful enough,....and the never ending war drum keeps beating. Please be clear here....I'm certainly not saying it's not ok to set goals and accomplish, and accumulate... (within reason...as long as Sacred Love is the guiding light we live by (much more about that to follow).....and basics like ....food, clothing, and shelter are equitably available for all).

And this War thing......

War against who?...against what?......hmmmm......could it be....ourselves?

And the lie that.... they are "different". "those people". Again, a seeming oxymoron.....for we Are each unique / different)....... and yet, each the same I.e, have the same needs.... for food, clothing, shelter, safety, "security", caring for and protecting our loved ones, feeling appreciated and worthwhile....

So yes......we've "been had, took, hoodwinked, bamboozled, led astray"......often by <u>bullies</u> ...who want to feel better about themselves, because of their own insecurities, so they convince themselves that the "differences" of race and ethnicity, gender, sexual orientation, class, animals, the earth, etc. ......are a threat, something to be "feared"...not loved...and even "less than, expendable, disposable". And then they use those lies to "elevate" themselves into positions of power, wealth, control, elitism, entitlement, authority / superiority...

and yet.....and yet......
this can also be viewed as ....

Men and Racism: The Healing Path

...a beautiful opportunity for connection......back "home" with ourselves.....and with that.... the desire to connect with others.

If we are willing to look deep within, beyond the surface lies, and many attractive distractions......we can use this awareness of the similarities as the ticket to freedom..... free from so much of the alienation that separates us from each other.....and most hurtfully.....from ourselves.

As Gabor Mate teaches us in his wonderful video, 'The Wisdom of Trauma'.....the greatest trauma is to be alienated from ourselves. And that stems from not being taught and modeled and cared for in a way that teaches us ….about All Life Being Sacred...and that each of us is Sacred....and Worthy of Love.

Story 2. **Fear of Women**. Wolfgang Lederer, wrote a book by this title. He told stories of how, from the cave people on....we have been in absolute awe of the amazing power of women....and their sacred connection to the earth, and the miracle of their life giving gift. So imagine our original male ancestors' fear, jealousy, insecurity etc. etc. of women, who had these mysterious powers, while men did not. Plus...and this is a big one....we have this amazing appendage between our legs, which, as any of us who have been injured there, well knows, definitely makes us feel Vulnerable...with a capital V. So we men found ways to help ourselves feel more secure by lording power over women, children, animals, the earth, each other.....all to prove our worth, and to feel more secure. Can you say ...suits of armor, and war, and competition to be number 1, at all costs no matter who we need to walk over to get there. **And can you say Racism, Sexism, Homophobia, Classism, Xenophobia...and on and on.... into infinity....all to feel more secure and to prove our importance / worth. Anyone we perceive as "different", or a "threat" to us, is automatically seen as the "enemy"**

I hope this is making sense...and not creating defensiveness. As that sure is not my intention here. My hope and wish and prayer...is to better help us understand the root causes of Racism etc, so that we can come back home to the beauty, and essence of who we truly are.......Sacred....Love. **And that Fear is a normal reaction.....to the lack of Love.....and simply a call to Loving Action.**

The question, is always....What would love do now?.... What is the most loving thing I can do for myself in this situation? And this, with the assurance...that loving action will always serve all best...even though it may not seem so initially. This is because loving action, is not always easy, takes a lot of courage, and some time takes time to see and feel the benefits.

Story 3. So Back to My Beloved Eileen. And **a story of Sacred, Love, Fear, and Being Vulnerable**.

Eileen was, and always will be, The Love of My Life. She passed five months ago, yet is still very much alive inside me, and always will be. While the pain often feels sooo deep....and wise one reminded me recently, that the deepness of the pain is a reflection of the deepness of the Love. And I will always feel so Blessed that she chose to Love me. She taught me how to Live Love......how to put the Sacred...and Love...into action....by offering me unconditional love and acceptance....so that, along with my own commitment to learn how to love me, and to share that love with all life, I finally felt brave enough to ….make a 100%, all out / all in commitment to 100% impeccable integrity with another being....by being.... yep......completely …..VULNERABLE!

So we were walking, hand in hand, as we often did, along the Tow Path next to the Canal.....when ...Love and Fear collided inside me.....and I collapsed crying on the ground. Eileen kneeled down and held me...gently asking...."What's wrong Roberto?" When I was able to collect myself....I shared with her, what I had never done with anyone else in my life (and I had been married, and had a step child, twice before).....that I was scared to death/life of losing her. If she drove back to Phily and got killed in an "accident"...I would be devastated...and didn't know if I could go on. That's when I knew...undeniably... felt it ...to the deepest core of me.......LOVE ....and the FEAR....of BEING ....VULNERABLE......YET WITH THE CERTAIN KNOWLEDGE THAT I HAD GIVEN MYSELF THE GREATEST GIFT IN THE WORLD...... LOVE..... Choosing to lovingly face the fear and risk of losing her...and the love we co-created, by our mutual spiritual commitment to ourselves and each other.....was and will always be....my single greatest accomplishment. Was I and Us...perfect? Hell / Heaven ...NO!.... There was a lot of rough going....especially since I am such an imperfect human.. with many, many imperfections. However, as they say.....we were and are perfect together.... because we did our best to honor our commitments to the ingredients of Love, (more about the ingredients of Love, a la bell hooks and her book 'All About Love' later in the All Life Is Sacred section.)

And when we fell short, because we are both very human...we did our best to prevent any drama..... and to acknowledge and understand what happened, what our plan to change was, and do our best to carry through. One of our many great gifts together. From choosing to commit to ourselves and each other...through FEAR, SACRED, LOVE, VUNERABILITY.

Stories 4 & 5

Holding myself in unconditional love and acceptance... holding the other in unconditional love and acceptance (with heartfelt thanks to one of my great teachers Dr. Amanda/Aminata Kemp). These two stories are about "broaching"..that is addressing sensitive, racially charged opportunities....through Fear, Sacred, Love and Vulnerability.

Story #4. The confederate flag.

While at a gathering at my beloved BFAM (brother from another mother, Johnny's), my beloved grandson Allistair, a dark skin African American, was walking by, at the precise moment Johnny's brother in law J. was removing his shirt, to reveal a tank top/muscle shirt, and confederate flag tattoo on his shoulder. I was stunned, walked out of the porch, to collect myself, and then remembered what Aminata had taught me. So I breathed, deeply....again, and again, and first held myself in unconditional love and acceptance, and then holding J. in unconditional love and acceptance, and then approached him. I did so respectfully, with a sense of curiosity about the origins of his tattoo, rather than attacking. He initially got a little defensive, assuring me that he wasn't racist, and that some of his best friends were Black. So I listened as he told me of his past growing up dirt poor in the South, and that the flag was a symbol of Southern pride...clearly of great importance to him. As I listened, and heard more, I was very touched by his rough life, and actually felt some tears. He did also seem to realize how the flag could be seen as hurtful to my grandson. Fear, Vulnerable, Sacred, Love.

Story 5. Black Lives Matter

While holding a large Black Lives Matter sign, and supporting a dear friend Linda R. who was doing her weekly one woman racial justice protest outside the front gates of Princeton University, a car drove by and the driver shouted "All lives matter, you racist!". To which I immediately, instinctively raised both hands and gave him two peace signs. He did a u turn and pulled up on the curve. I said to myself .. "Oh boy, here we go". However, another miracle happened....I was again somehow able to summon unconditional acceptance and love for myself, and then for him...As we both approached each other, I said to him....You know....you are right...."All Lives Do Matter", and "We're all being screwed by the wealthy one percent at the top who have all the power, wealth and control. That stopped him, he said, "Yeah you're right!". He did agree that Black folks had it rougher from the beginning and now. We then both shared our common backgrounds as veterans, and he ends up inviting me to go on the nearby tours he led of Washington's battle grounds. Fear, Sacred, Love, Vulnerability.

We men

Since the beginning of time
we men have tried,
to be in charge,
while many have died.

We thought it our duty
our right, our job,
to run the show,
and even to rob

others of their rights,
a negative effect
preventing so many
of freedom and respect

If we're in control
have the power and the toys
we think we're men
but really only boys.

For the values often taught
can fool us into thinkin
that the answer lies outside us,
like money, and like drinkin.....

Then finally one day,
we awaken from the haze,
and we realize we've been duped,
not been our finest days.

And hopefully what we get,
is the wisdom from above,
That what we've always needed.....
is a heart that's filled with love.

FRS/10/2/19

The following is a slide from a Boys to Men Multicultural Counseling presentation.

Boys to Men: *Beyond Diversity**

This is about how to explore <u>being male in a toxic, macho world</u>, and how to be healthy and happy in our *white supremacy culture.*
The goal is to provide safe, supportive encouragement as we uncover the obstacles and strategies to finally coming back home..........to ourselves.

Initial Considerations/Generalities (disclaimer/a lot of generalities about men/important to be sensitive to the many differences and distinctions, among and within families/cultures:

1. <u>This is not about blaming males</u>. Quite the contrary, it's about empowering boys and men to make a loving difference in our own lives and the world around us.

2. While we males have made and continue to make so many positive contributions to our world, we are also, unfortunately, responsible for contributing to the vast majority of violence, oppression and destruction.

3. Because of the ever increasing amount of anxiety, fear, depression, stress related problems, more and more males are <u>feeling alienated from ourselves</u> and each other.

4. Increasing numbers of males coming from multicultural backgrounds continue to present unique opportunities and challenges for re-creating a society where "everyone" feels welcome....where "We the People" truly means "All" the People.

5. There are important similarities and distinctions related to cultural, family, class, economic, gender backgrounds. So this is about celebrating our cultures, uniqueness/ commonalities....and about coming home to ourselves.

6. We all have privilege (ex.,food clothing shelter, healthcare, some used for positive, some not/i.e., movies Blind Side (positive examples), Avatar (negative exampls)/ biases/from growing up in culture of fear/ some ex's of my own privilege are being white hetero male/able bodied, educated taught some traditional biases i.e., women should stay at home, hetero men are macho/gay men are effeminate.

7. Depending upon family and culture, messages little boys learn about <u>how</u> to be men often include qualities like being tough, competitive, not crying or <u>being vulnerable</u>, always being in control,

powerful, successful..<u>self worth is based on externals</u>/ accomplishments/ goals, material wealth/ basic human needs for comfort, safety, nurturance, respect, acceptance, connection are all too often overlooked.

8. Depending on family and culture, initiation rites of passage/traditions/rituals/mentor- ing/role models for entering into manhood have been lost in this highly industrialized world – often resulting in the emphasis on competitive sports, gangs, overindulgence in aod.. any of the miriad of distractions, gambling, internet etc., just to feel ok/adequate as a man.

9. Dark side of competition- to be #1 at all costs, no matter who you need to walk over to get there, becomes a value....no matter what, it's rarely good enough<u>/ "I'm not good enough"</u>/ which often leads to great stress/anxiety/depression/self-harm.

10. Some cultures allow the expression of emotions and feelings ex. anger/excitement, however, all too often, the emphasis is on being stoic and in control (for fear of being perceived as weak and vulnerable. It has been said that <u>deep anger</u>, sadness, and lack of ability/permission to cry and grieve loss is the main contributor to serious <u>dis</u>-eases amongst men. Rage work and grief work in supportive counseling can be so helpful.

11. Domination over others and the environment is seen as high value.....however we <u>actually</u> have little control over either.....which then becomes a real dilemma and significantly problematic.

12. There is often a stigma attached to asking for help – seen as shameful, <u>fear of being viewed as or feeling weak/vulnerable</u> (shame attached to fear of being overwhelmed by emotions....<u>so deny have any...in order to maintain semblance of control and familiarity</u>).

13. Feeling inadequate due to moving to new culture and not fitting in, language barriers, financial challenges, unemployment, not being adequate provider, history of trauma/abuse, imprisonment, not meeting social stereotypes of manliness, separation/divorce, longing for acceptance, safe touch, affection.. all contribute to a sense of desperation and fear.

14. Issues like white heterosexual male supremacy/privilege are rarely if ever acknowledged, leading to confusion and self-blame for those who don't fit into that group.

15. Since men are not allowed to voice fear, even though we all live in cultures of fear- given the economy, poverty, terrorism, global warming,

violence, wars...<u>anxiety and depression related disease are rampant</u>. (*Parts of the above were taken from a keynote presentation by R. Schiraldi at the annual NJ Multicultural Counseling Association conference).

<u>Self-Care Strategies</u>*(excerpts from R. Schiraldi White Privilege Journal article):

<u>1. Be willing to continue uncovering our own privilege and biases as we continue to work on enhancing our self-awareness so we can become more comfortable talking about diversity and privilege with peers and clients. For example, keep a white privilege journal of reminders and new learnings about our biases and privilege, as we continue to affirm and acknowledge our efforts.</u>

2. Remember that this can be hard work, that takes <u>a lot of empathy and courage.</u> So it's important <u>to be gentle and compassionate</u> with ourselves. For example, taking lots of deep breaths, and giving ourselves lots of "atta girls" and "atta boys", and breathing in comfort and words of nurturance and encouragement.

3. <u>Look for allies</u>. This work is too hard to do alone. Finding others who feel the same way, can be so helpful in preventing burnout and discouragement.

While it might be difficult to find allies to do this work – it is essential. If there are no internal work related groups, there are usually some organizations in the community which address privilege and multicultural concerns.

4. Believe in ourselves. We can make a difference, one little step at a time, one person at a time. Along with keeping a journal of our efforts, graphing progress can also be useful in keeping our challenges and progress in perspective.

5. Be prepared for negative reactions to our efforts and try not to take them personally. The issue of privilege brings up a wide range of feelings in people. Journaling, mindfulness meditation and <u>non-violent communication</u> can be very helpful in providing self-comfort as we <u>acknowledge our feelings and needs and other's feelings and needs</u>. When people are <u>feeling unsafe and threatened</u>, they will often react negatively.

6. Consider letting others know we are open to discussing these issues. For example, have pictures, books and artifacts in our offices that help make it clear of our sensitivity to multicultural issues. When appropriate, share our interest in discussing these issues.

7. Be respectfully curious about others' family, name, religious, cultural traditions.....***don't assume that because we have similarities our experience and feelings are the same.

8. Seek to learn more of family/cultural messages we and other boys learn about becoming men, initiation rituals, role models.

9. Making empathetic connection with self and (when ready, with others), ex. what is the fundamental pain I'm struggling with, and seeking support to address the pain.

10. Acknowledging resistance..... look at why, and what might help to feel safer in further exploring.

11. Acknowledge how difficult it is to talk about pain/normalize physical symptoms, irritability, sleep, headaches, risky behavior, all common when feel alienated. Takes courage to be willing to learn compassion/healing...however if don't, will continue to feel less than whole.

12. Be real – pretending we are perfect is crazy, and a set up for tons more stress..this is tough work....we are not going to do this perfectly, and we need to be kind with ourselves as we pull back the layers.

13. Reassure ourselves about learning how to deal with feeling different and disconnected, for example through seeking support groups that address related issues such as multi-cultural concerns, <u>being male/ teaching ourselves and our young men how to be gentle with ourselves and our world</u> (Kivel, 1993), healthy relationships, and managing emotions, with approaches such as dialectical behavior therapy (Marra, 2005), mindfulness meditation (Kabat-Zinn,1990),and learning how to meet basic human needs such as acceptance, connection, empathy, through learning the <u>art of non-violent communication</u> (Rosenberg, 2003). Each of these approaches teach specific tools about learning self-respect and emotional management with ourselves and empathetic connection with others. Learn language of feelings, bodily sensations (tension etc.), emotional regulation log, practice sharing feelings. Compassion Focused Therapy/ compassionate self talk practice. Not taking away defenses until have something to replace them with, ex. NVC, needs, compassionate self-talk/actions.

14. <u>Remind ourselves about our internal worth, rather than looking for external validation</u> (G. Schiraldi, 2001). Look for opportunites to celebrate our uniqueness, and of our interconnectedness to

all other living things. For example use the crystal picture found in chapter four of Schiraldi's Self-esteem Workbook, to demonstrate how <u>our essence is worthiness</u>, however we've often been taught a belief system that focuses on <u>materialism and external validation</u>. Look for opportunities to encourage a radical paradigm shift to believing in self-worth, <u>as a given / our essence</u>, not having to earn it, viewing accomplishments as "gravy". Have fun and be willing to consider creative approaches. As difficult as this work can be, it is very exciting, enriching and rewarding . Keep a list of new approaches, strategies, tools gained from interactions, readings and attending trainings – especially those that help us laugh, as keeping a good sense of humor is so important in doing this work.

15. And finally, if you really want to go for it - Consider <u>setting an intention, making a pledge, writing a commitment letter </u>(whatever works), to address cultural competency and privilege in an ongoing way, personally and professionally, for example, discussing it with friends and colleagues, getting more training, attending the annual White Privilege Conference, asking for support from your boss to have staff training, joining work groups which address cultural competence, and maybe even (take a deep breath) being willing to give up

some of our privilege, temporarily or permanently, in whatever way feels right to us for example, eat less or periodically fasting, donate time or money to charitable and social justice causes, conserve energy or recycle. _**Please remember to be gentle and kind with our precious selves, and all life, The Greatest Strength Is True Gentleness-Lakota, Mitakuye Oyasin**_

Kabat-Zinn, J. (1990). *Full Catastrophe living.* New York: Delta Publishing.
Kivel, P. (2002). Updrooting racism: how white people can work for racial justice, New Society Publishers.
Marra, T. (2005). *Dialectical Behavior Therapy in Private Practice: A practical and comprehensive guide.* Oakland: New Harbinger Press.
McIntosh, P. (1988). *White privilege and correspondences through work in women's studies (Working Paper No. 189).* Wellesley, MA: Wellesley College, Center for Research on Women.

Rosenberg, M.B., (2003). *Nonviolent communication:A language of life.*California:PuddleDancer Press.
Schiraldi, G.R., (2001). *The Self-Esteem Workbook.* Oakland: New Harbinger Publications.
*Schiraldi, R., (2011). *White Man on the REZ,* White Privilege Journal, University of Colorado, Vol. II, Issue I.

My Sacred Feminine

Alive
In me
Is the
Miracle
Of Life.
Continually
Being reborn
Renewing
Recreating
Nurturing
Sustaining
Energy.
Compassionate
Soothing
Comfort.
Forgiving
Humility.
Healing
The wounds
Deep
To the core
Of Mother Earth
Of me.
A gentle
Salve

Washing over
The scars.
Joyfully
Pronouncing
Freedom.   FRS 12/06

Mental health and wealthy white hetero male supremacy culture....the Dilemma.

Mental health can be viewed as a result of experiencing a balanced life...emotionally, mentally, spiritually, physically, socially.....and a belief in one's self-worth. Few of us have been raised by perfect parents. So it is up to each of us to choose a belief in and acceptance of our core worth. If our worth depends on external circumstances or influences, such as our accomplishments, job, material wealth, and approval of others, our sense of self-worth will be transitory at best. It is up to us <u>to choose,</u> <u>to decide</u> that our essence is whole and worthwhile...and that all other aspects like feelings, ailments, highs and lows, are just part of life, and don't really affect our true self-worth.

While there are many contributing factors to mental illness, I believe that one huge cause has rarely been touched on....the culture of wealthy white heterosexual male supremacy. This cultural value system of wealthy white hetero male supremacy which most of us have been raised in, is at this very moment in history being severely tested, especially given the combination of the nationwide protests of continual murders of Black men, and the life threatening corona virus. I don't believe there is any coincidence that these two major life-altering events are simultaneously occurring. I believe that both of these events are happening at this exact moment in time, to hopefully awaken our sense of outrage at the life threatening value system of wealthy white heterosexual male supremacy, which is the root cause of most, if not all of the dysfunctionality and pain that plagues our society. This system is a major cause of mental disease, disorders...because we are dwelling in a disordered culture of fear, where our very humanity, and sense of self-worth is never good enough. We, (especially we men), have been taught that a sense of our okay'ness depends on so many things outside of our control. This leads to incredible stress related illness and mistrust of ourselves and each other...especially of other men of color, who have been perceived as a huge threat since the beginning of this country. This is because, while there were some good "intentions", the country was never founded on true equality and justice for all....only for the few wealthy white heterosexual men (the Constitution excluded Native People and Women, and claimed African American males as 3/5ths of a person, while the Declaration of Independence declared Native People as "savages". Mental health largely depends on living with impeccable integrity, compassion and humility for ourselves and all others. These qualities have never been more questioned that at this present moment.

So let us explore what the words …..wealthy…white….
heterosexual….male…..supremacy imply…so we can better
understand the impact on our lives.

wealthy white hetero male supremacy terms defined:

wealthy – land owning, material wealth, has been a key value for power since the beginning of this country, and a symbol of success, achievement, **worth,** prestigious.

white – white is right, at least in this culture, is key, given that the white supremacy cultural system of racism was started and still run by predominantly white males, white seen as good, Black and other colors not as good, or even bad, evil, certainly at least, less than, not as smart.

hetero – next key ingredient…..almost synonymous with masculinity, fear of being seen as weak/gay/effeminate/ vulnerable, don't share feelings, rational, macho --Fear of Women, penis envy, women secondary.

male is the key ingredient…has been since the beginning of time….male cultural values…competition to be #1 at all costs, power, material wealth, control, success, rational/unemotional, aggressive,

supremacy – to be in charge, at the top, the most important, to have all the power, superior/supreme, #1, to aspire to, entitled, elite, rulers of the land, animals, all others...(almost god like).....

......while most self-secure men would not claim the above as their core values....one cannot deny that these were the core values and still are in terms of who has the power in this country.

The above warrants a lot more examination.....because if the roots are still poisoned, so will the efforts towards growth be poisoned. I hope that at least some of this essay touches you and supports you in understanding how peeling away the layers of our white heterosexual male supremacy culture will begin to free us to work towards healthier mental health. Balanced emotional, mental, physical, spiritual, social health demands courageous truth telling and compassionate action for us to feel secure inside, and to make things right.

I offer these words below for comfort and encouragement for each of us......

Breathing in, breathing out...

I am whole and worthy, exactly the way I am

I serve all others with my whole heart and my whole spirit.

May it be so.

(For a little levity)

Monkey Suit.... Defined

There is much "rich"
historical background....
basically, a term
I learned and used
as a kid
in Brooklyn and East Rockaway...
since I hated being forced,
to dress in suit and tie...

very uncomfortable
very formal
very stiff...

like a trained monkey....

to act
to walk
to talk

an acceptable,
prescribed way...

very homogeneous
constricted
stiff

reserved
inauthentic
politically correct

properly assimilated.....

hard.... to... breathe.

A uniform
for
the power elite.....

Not to be trusted.

When the tie is loosened,
breathing improves...

and Love can flow.

2017 / 2020

P.S. If this speaks to you about the gift of loosening up, if the situation allows....great....if not, hope I didn't offend.

Part II

Living Change.....
Through Loving Action

The following is an outline for a proposed organization in support of men coming together to support each other in learning, growing, healing, to become better men, and support our boys in becoming healthy men, in service of a better world.

Men for Racial Healing and Justice

Men for Racial Healing and Justice is an organization of men who are concerned about addressing racial healing and justice for ourselves and the world around us. While we males have made and continue to make so many positive contributions to our world, we are also, unfortunately, responsible for contributing to the vast majority of violence, oppression and destruction. We believe that effective racial healing and justice must begin with us men holding ourselves and each other accountable and standing together to insist on racial healing and justice for us all. Safe, supportive environments are provided in which to do this most challenging and often painful work. With the support of other safe, strong, caring men, together we can heal ourselves and the world.

Our mission is to continue to deepen our understanding of the white heterosexual male supremacy value system / how it has hurt us and our common humanity, and pitted us against ourselves and each other. This is the underlying, belief which drives our efforts. Fearlessly speaking this truth with love and action is our commitment.

<u>Advocacy/Consultation/Individual and small group support</u>

Providing individual and group support for stepping up and speaking out about racial justice. Exploring efforts to effectively interact with individuals and systems. Providing support for healing from racial trauma, and other related concerns.

<u>Programs</u>

A variety of programs are designed for organizations and institutions to explore the roots of internal and external racism and the ways to change ourselves and the world in ending racism.

<u>Two Day - Four Day Intensive Gatherings.</u> "The Many Myths of Being Male: The Keys to Ending Racism".

These programs will allow for in-depth internal work and creating relational and team building opportunities as we peel away the lies that have oppressed us and hurt our common humanity. We will come away feeling refreshed and invigorated in our commitment to healing ourselves and ending racism.

Spiritual Healing Ceremony
Meditation and prayer circles
Sweat Lodges
Drumming Circles
Embracing the Quiet Within
Learning how to be compassionate with ourselves, each other, and our Earth Mother.

Program topics to include:
Growing up male
Stereotypes/gender role conditioning
Competition to be #1 at all costs (no matter who you need to walk over)
Power, wealth, control
Real strength / gentleness
Fear of Women (based on book of the same name, by Wolfgang Lederer)
Fear of being vulnerable / the importance of honoring our emotions / how to be safe doing so (based on the work of Brene Browne)
The Enemy
The Warrior
The King
Fatherhood
Friendship/brotherhood
Healthy relationships
Healthy Sexuality

Celebrating Maleness
Being impeccable with our word
Our Sacred Path
Our Sacred Nature
Wealthy white hetero male supremacy cultural values
LOVE, what it really means, how racism is the lack of love, and how real love for ourselves and all life, is the antidote (based on book 'All About Love', by bell hooks).

Maximum Security

Maximum Security

The Ultimate Sham!

Unless

It's internally focused.

Then it's cool.

FRS

Double Jeopardy!

Implications of Healing From Trauma
For Native, African American and Other
Veterans Of Color

Many Veterans of Color are very proud of their military service. And they certainly deserve to be. However, for many, sharing stories and feelings about the traumatic effects of military service on themselves and other veterans of color can be a very challenging problem. The following statement, and article which follows, summarize some of the sentiments I have heard from fellow veterans of color since serving in the military back in 1969 - "I'm supposed to feel proud of my service, and I am. However, I am part of a war machine that massacred and enslaved my ancestors, and continues to enforce unjust policies here and around the world, often against other poor people of color".

This can be a painful, and morally conflicting dilemma.

I am a white male Vietnam Era veteran. So where do I get off writing about trauma for veterans of color? As a white male veteran, I feel it is my responsibility to speak on this topic, since it is still predominantly white males who hold the power in our government and in the military. And since the beginning of our country to the present, it is primarily us white males who have perpetrated the violence against people of color, women, children, animals, the environment.

After my infantry training, and combat medic training, I was trained to be a neuro-psychiatric specialist, serving in an intensive care psychiatric unite of an army hospital. While there I was
honored to work with and learn from many veterans of color, and experienced first hand the often very excruciating difficulties they had to deal with, and are still having to deal with.

During most of my career as a trauma therapist and racial justice advocate, I have been taught by many of my sister and brother veterans of color about the unique challenges facing them and that all too often go unspoken, due to rage, guilt, shame, fear. So it my hope to shed some additional light on how necessary it is for us to promote preventative training about these issues and to ensure improved services to our veterans of color as a step in healing the wounds.

Post-traumatic stress disorder, PTSD, was not a term we used back in 1969. As the Vietnam "conflict" was winding down, many combat veterans were hospitalized in intensive care psychiatric units as a result of the horrific atrocities they experienced.

Reflecting back now on some of the veterans of color I worked with on the wards and alcohol and other drug rehab wards, I am struck by the almost impossible situation they found themselves in. They were not treated like full-fledged citizens, yet expected to put their lives on the line, and go kill other people of color who they had no grudge with.

This brings to mind Muhammad Ali – brave enough to make a stand, yet having to sacrifice the best years of his career.

Many of our veterans in Vietnam were ordered to commit unthinkable acts against innocent men, women and children, or face being shot themselves, court-martialed and long term imprisonment for disobeying orders. Then they return home from the trauma of war, to their country, and were not welcomed as heroes, but were met with protests and criticism. And veterans of color were thrust back into a country caught in the middle of heightened awareness of painful civil rights injustices, compounded by intergenerational racial trauma, which is only now beginning to get the attention it deserves. They were and still are, often faced with "red-lining" practices to prevent buying homes in certain neighborhoods, as well as many other discriminatory practices like when applying for jobs and bank loans, and racial profiling.

I was drafted while serving as a Teacher Corps Intern, as part of the government's anti-poverty programs. Along with some Vista volunteers we would apply to rent or buy from landlords or real estate agents who had turned down African American Veterans. Attorneys would then step in to help enforce fair housing practices.

Many Native American veterans experienced shame for feeling like they betrayed their people by serving in the military. Their ancestors had been raped and massacred by U.S. soldiers, and survivors forced to travel to desolate reservations where poverty is rampant to this day. Alcoholism, other drug addiction, extreme domestic violence and abuse, suicide, many physical diseases and ailments, can be directly linked to the injustices which have never been rectified

I worked with a veteran on the psych ward who had been a member of the Black Panthers when he was drafted. As with many of us, being faced with jail or having to leave the country, he chose to serve. Being torn inside about the killing he participated in while in Vietnam, as a Black man / Black Panther, led him to be suicidal/homicidal and eventually to the psych. ward. Being faced with returning to his city feeling unwelcome, even a traitor to his people, was untenable to him. This was just one of many similar stories of the painful dilemmas facing African American and other veterans of color. It is not only the aforementioned injustices, but the often unspoken guilt and shame which is responsible for many suicides, alcoholism, other drug addiction, domestic violence and abuse and many physical diseases and ailments.

A few years back I attended a powerful meeting in NYC of a group called Intersections which hosted small group discussions to address these stories to heal the guilt and shame.

War is hell for all soldiers. However, it is certainly a lot clearer to me now, that as a white heterosexual male, I didn't have to deal with the same level of traumatic stress that my Native, Black, Latino, and other veterans of color, women and gay sister and brother veterans did and still do. This includes harassing, oppressive treatment by fellow soldiers and those in charge.

It is way past time that we address the impact that military service can have on the mental, emotional and physical health of our veterans of color. Training for government, military leaders, mental health and other health care professionals is essential about how to best prevent and reduce levels of trauma, and how to acknowledge and tend to their treatment needs.

Thank you.

Roberto Schiraldi

("In the Second World War over a million African-Americans fought for freedom and democracy - in an army, that was strictly segregated by race. These African-American GIs fought to liberate Germany from Nazi rule, where racism had reached a dimension that was unfathomable.
Narrated by Academy-Award winner Cuba Gooding, Jr. and featuring interviews with former Secretary of State General Colin Powell and Congressman John Lewis, this is the remarkable story of how World War II and its aftermath played a huge role in the Civil Rights Movement. It's a story told through the powerful recollections of veterans like Charles Evers, brother of slain Civil Rights icon Medgar Evers or Tuskegee ace pilot Roscoe Brown. **From the beginning, black soldiers felt the absurdity of being asked to fight for freedom while being denied it in their own army."**
(From the highly recommended documentary "Breath of Freedom", available on internet / Vimeo / Broadway on Demand, December 2014.))

With gratitude for the aforementioned to my Red Fox bench buddy Lorenzo, and for honoring me with some of your family's sacred history.

Please see the important book titled 'Unconventional Combat: Intersectional Action In The Veterans' Peace Movement', by Michael Messner, for an important examination of how the intersectional marginalization of veterans from different gender, race, class, sexual orientation so painfully oppresses those individuals and groups.

Also please see a touching and inspirational little book titled, 'America Needs A Woman President', by Brett Bevell, drawings by Eben Dodd, which demonstrates how hugely important it is for us to have more Women in leadership positions, especially Black, Indigenous, LGBTQ2S, and Women of Color who are wise, compassionate, and courageous.

Thank you Katherine Hernandez for your inspirational leadership, and my other sisters and brothers in the "Be The Change" book club of Veterans For Peace, for your growing efforts at addressing sexism, racism, homophobia/transphobia, classism in our militaristic society and within Veterans For Peace, and for your love and support of my own personal work in these areas.

Dog Soldier

I stake my life.
This is where I make my stand.
I pledge forever
To give my Life
So that
All my relatives may live.
I pledge to Live
In each moment
With impeccable integrity
With courage
With compassion
With love
With joy
In peace
And never give away my power
To negativity and fear.
 To offer kindness
And nurturance.
To protect
 The children
And the old ones
The poor
The animals
Our Earth Mother
With my Life.

I offer my Blood
Which deepens
My pledge
To All Things.
THIS   Is My Word.   FRS 1/07

Men Stepping Up

Veterans For Peace.....Still a white male dominated organization?

The intersection with white supremacy cultural values.

Stepping Up.....

……… Stepping Over....

…... Stepping Back

A delicate dance

Speaking out...yes.....listening more....yes. Supporting and encouraging those who have been marginalized to speak out and to have their voices heard...and to move into shared leadership roles, are certainly worthy goals for Veterans For Peace. I know VFP has been conscientiously working to address these concerns, i.e., increasing veterans of color, women veterans and allies as members and in leadership roles. To be a more inclusive organization, this work is essential.

To attract and sustain a widely diverse membership, which warmly welcomes and values all, how we interact with each other is an issue which warrants continual care. Knowing when to speak up, and when to defer to the voices of folks who have been marginalized, requires much compassion, sensitivity and patience. All too often, I, as a white hetero male who is passionate about equity for all, can step up and speak out....and then step over / talk over others, with emphatic, long-winded answers, thus doing the very thing that I am so passionate about changing. Men having all the answers is one symptom of white supremacy. I sure don't have all the answers, and need to know when to ask for guidance from those who have been so hurt by this type of behavior.

Ahh, passion, commitment...patience and listening.... really listening. ....such important ingredients for fairness and trust building....for us all.....and especially for us men. I have been trained to compete for number one, and to make my voice be heard. And I do my best to re-member that might doesn't make right, and being number one, can sure be a lonely and dehumanizing place to be. I joined VFP because I learned of our commitment to addressing racism, patriarchy/sexism, homophobia, militarism, and discrimination and oppression of all the other marginalized people, and the desecration of our Earth Mother. So I offer what follows as a partial antidote.

Proposal for Veterans For Peace white supremacy culture statement resolution

Introduction

Fortunately, we men have been responsible for wonderful contributions to the planet. Unfortunately we have also been responsible for the majority of violence, oppression and other atrocities perpetrated on our planet. And also, unfortunately, more and more women seem to be emulating these cycles of violence.

Veterans For Peace has courageously and compassionately taken a lead role in promoting peaceful healing for our planet. One of the ways we can continue to do so is by speaking out against the culture of war value system of our country, a <u>white supremacy culture</u>. Since white men were, and still are predominantly the ones in power, it is more aptly described as a wealthy, white, heterosexual, male, supremacy culture. This "culture" cleverly manipulates and pits all marginalized groups against each other. This system – a hierarchy of human value where inferiority and superiority are based primarily on race, is the very same system which founded this country and continues to run it. It is based on the Doctrine of Discovery and Manifest Destiny which saw indigenous people and enslaved Africans as sub-human savages, and God's anointed ones as the rightful owners of women and the land. It justified the genocide of millions of indigenous people, the murder and horrific treatment of enslaved Africans. It continues to justify a Jim Crow prison system, and inhumane treatment of refugees and immigrants of color. Basically anyone who is not a wealthy white heterosexual male is fighting for a very "unequal" piece of the pie.

This is a set up for violence and war.

Veterans For Peace continues to take stands for true equality, liberty, and justice for <u>all</u>, and the sacredness of <u>all life</u> – including women, children, animals and the environment.

Wealthy white heterosexual males are predominantly the ones in power in our government and military. Therefore, it is important that we white, heterosexual males be the ones to continue to speak out and support efforts of peace related coalitions, especially those efforts addressing equality and racial justice for <u>all</u> people, and humane treatment of women, children, animals and the environment.

As our foremothers and forefathers have taught us - "When we are not part of the solution, we part of the problem".

Below is a statement addressing the culture of white supremacy, which is submitted in hopes that VFP will consider adopting something similar. It is based on a resolution adopted in 2018 by the National Education Association, and similar statement adopted by the New Jersey Counseling Association in 2020.

Draft for Consideration

Statement on white supremacy culture

Veterans For Peace believes that, in order to achieve racial and social justice, we must acknowledge the existence of White supremacy culture as a primary root cause of violence and war, both internally and externally, of institutional racism, structural racism, White privilege..and most especially intergenerational and daily racial trauma. Additionally, we believe that the norms, standards, and organizational structures manifested in White supremacy culture perpetually exploit and oppress people of color and serve as detriments to racial justice, and healing from racial trauma. Further, the invisible racial benefits of White privilege, which are automatically conferred irrespective of wealth, gender, and other factors, severely limit opportunities for people of color and impede full achievement of racial and social justice,

which are necessary for healing associated with racial trauma.

We consider it essential to acknowledge the history of the displacement of the Indigenous People in this country, the legacies of the enslavement of Africans and their descendants, the **treatment of all other marginalized groups, and to celebrate their accomplishments.**

Therefore, Veterans For Peace will actively advocate for strategies fostering the exposure of and eradication of institutional racism and White privilege perpetuated by White supremacy culture, starting with ourselves and our organization. We believe this will support our mission of ensuring racial justice for *all* people, especially those who are severely hurt by the ongoing disease of white supremacy/racism.*

*Adapted from the National Educational Association Report 2017-18
(New Resolution, White Supremacy Culture, July 2018).

References:
Chauduri, D. Santiago-Rivera, A. Tlunusta Garrett, M., Counseling and Diversity, Cenagage Learning, Belmont, CA, 2012.
DiAngelo, R., *No, I Won't Stop Saying White Supremacy: Naming white supremacy shifts the locus of the problem*

to white people, white people, where it belongs, Good Men's Project Newsletter, April 2017.

Hemmings, C., Evans, A., *Identifying and Treating Race-Based Trauma in Counseling,* (2018),Journal of Multicultural Counseling and Development, Vol. 46, Issue I20-39IV..

Liu, W.M., *White Male Power and Privilege: The Relationship Between White Supremacy and Social Class,* (2017), Journal of Counseling Psychology, Vol. 64,.

McIntosh, P., *White Privilege: Unpacking the Invisible Knapsack,* Wellesley College Center for Research on Women, Wellesley, MA, 1989.

Messner, Michael, Unconventional Combat: Intersectional Action in the Veteran's Peace Movement, Oxford University Press, New York, NY, 2021.

Okun, T., *White Supremacy Culture,* Notes from People's Institute for Survival and Beyond Workshop, Oakland, CA, spring 1999.

Smith, C., The cost of privilege: Taking on the system of White Supremacy and racism. Fayetteville, NC: Camino Press., 2007.

(This is a proposal I submitted to support caring folks who really want to make a difference in improving and helping to make the efforts of VFP more effective / affective) .

The following 6 poems and 1 song about climate crisis and militarism, were recently written in response to a request from Carol Rewart Trainer, a valued friend and fellow member of the Be The Change Book Club of Veterans for Peace, for her Veterans for Peace pod cast. Thank you Carol.

Climate Crisis and militarism

doctrine of discovery
god's anointed ones
manifest destiny, desecration of the land
for power, wealth and fun?

The original Native Ones
Honoring Earth Mother
military might
to control the other.

People of color, animals, trees
pollution of water and air
brings us to our knees.

All Life Is Sacred
The mountains, rivers, streams
military power
isn't what it seems.

If we choose to use that power
for re-creating and preserving life,
and check the arrogant greed,
senseless competition, bickering, strife.

~

Soo.....
To live in harmony with <u>All</u> life
Saving our children from the edge
This is the moment in time
To make a life-long pledge

To each do our part
to replace each gun with flower
It's up to us... right now
To make <u>this</u> our finest hour.

Mitakuye Oyasin...We are All in this Together.

~

Responsible?

Responsible to all of nature
Responsible to the land
Responsible to the animals
Responsible to the ocean
Responsible to the air we breathe...

Responsible to do our duty...

Responsible to kill ... anything in our way.

~

The Web of Life

Quantum mechanics
molecules and cells
all life connected
like complementary bells

A patriarchal system
wealthy men at the top
is a huge, huge set-up
for a very painful flop.

No better, no worse
no need to fear

with equity for All
no sittin in the rear.

Together with all our allies
we can heal our mother earth
So let us heal our mother earth
let us heal our mother earth.

And then, and then...
with all our relatives, connected
we can heal the strife,
and then we can dance,
the web of life.

we can let the dancing begin.

~

Coming Together

What if
we veterans
and all those
currently serving,

Decided to make a start...
and agreed to come together
in unity... one mind, one heart.

Honoring the planet
And all her precious gifts
Resolved to mend the fences
And heal the many rifts.

Military might
military right
in respecting all life
we light the darkest night.

Military might
military right
in respecting all life
we light the darkest night.

~

Poem of Gratitude and Hope

Our mother earth is hurting
Will we heed her urgent call?
We of the military
Will we walk proud and tall?

Will we speak to those in power
To be protectors of the land?
We can't do this alone
World-wide joining, hand in hand

With integrity and humility
Being courageous and kind
With compassion and strength
With our heart and with our mind.

So much beauty on our planet
So much beauty, so much worth
If we choose to soften, not harden
Tend and grow our lovely garden.

Walking gently on the earth
Deep respect for all beings.
May we each walk in beauty
Truly grateful for all things.

Pilamaye, Wopila.

~

Revitalizing the Land

Agent orange, all the bombs
On pristine islands, mountains and farms
The run offs, pollution, poison of young
Now we make it right, reparations everyone

Truth and conciliation for the land
the animals, the people's wounds
asking forgiveness on our knees
as we tend to the gardens and plant new trees.

Holding ourselves responsible
cleaning up our mess
holding our leaders responsible
cleaning up their mess

with love and respect
we <u>can</u> do this

responsible military
we <u>can</u> do this

responsible military
we <u>can</u> do this.

Will We?

The following is adapted from a Quaker sweat lodge song:

The Earth Is Our Mother

The Earth is Our Mother
We must take care of her.

The Ocean is our Relative
We must take care of her.

The Sky is our Father
we must take care of him.

The Mountains are our Relatives
we must take care of them

The Trees are our Relatives
we must take care of them.

The Animals are our Relatives
we must take care of them.

The Earth is our Mother
we must take care of her.

Pilamaya, Wopila

Recommendations For Police Reform to the New Jersey Attorney General

(Most of the following was offered as part of a multi-organizational proposal submitted to the New Jersey Attorney General)

Toward a Force of Peace Officers
Peace Begets Peace......Violence Begets Violence

Introduction / Background

Police Departments across New Jersey are filled with courageous, caring, dedicated, "public servants" who are committed to upholding our country's and our communities' highest ideals of a peaceful and harmonious co-existence, where all citizens are treated equally and fairly, with utmost regard for their safety, well-being and dignity. However, unfortunately, our police system is a reflection of a flawed system of government...which founded this country and still runs it.... a system steeped in a culture of white supremacy values. This system is set up to protect a few bad apples who do harm to the trust of the police as a whole. The term white supremacy culture is increasingly being used to describe institutionalized racism, (i.e, National Education Association, New Jersey Counseling Association, please see attached statements from those organizations and proposed statement for New Jersey Police Departments(1.)). **This culture of white supremacy is a system of power, wealth and control, where a predominately few wealthy white heterosexual male, "land owners" are in power - a system where inferiority and superiority are based primarily on race. In this system, all the racial groups, and all the other groups - women, LGBTQ, class, disabilities, and others, <u>including us white heterosexual men</u>, are cleverly manipulated and pitted against each other, each fighting for a "little" / "equitable" piece of the pie.** Native, Black people and women have always been marginalized, since the constitution intentionally excludes mention of Native people and Women, and counts Black people as 3/5's human. This systemic racism is deeply embedded in all of our institutions – our government, courts, schools, churches, banks, corporations...and police departments. This culture of white supremacy **hurts us all.....hurts our common humanity**. It needs to be exposed for what it is....a lie..... perpetrated for power, wealth and control, by a selfish, scared few. So to be more accurate it should be called wealthy, white, heterosexual male supremacy culture.

This is not about guilt or blaming, since we haven't been taught about the accurate history. However it is about holding ourselves, each other, and those in power accountable for sustainable change. Until our accurate history is clearly acknowledged and made right, any effort and hope for a just police system, which satisfactorily addresses use of force concerns, will surely fall short.

The following are proposals which strongly and compassionately support the reformulation of a Peace Force. (Since the first recommendation is the most controversial, explanations are indicated):

<u>Given:</u>
- The unresolved foundational history of racism in this country, is embedded in all of our systems.
- That police forces were originally established to recapture, imprison, and then return runaway slaves to their "owners". (Consider the implications of this statement for today's policing)
- That police departments are part of a larger system which was set up to empower wealthy, white land owning males, at the expense of Black people, Native people, and all other marginalized groups
- The continually occurring and worsening atrocities against Black people, and all other marginalized groups
- The increasing disapproval throughout the country of the police treatment of Black people
- The deleterious relationships, between FOP/DAs which prevent weeding out the "bad cops"

-That the "blue wall of silence" is reinforced by the aforementioned relationships between the FOP/DAs

Therefore:
1. The best chance for a true Peace Force of trusted public servants, is by "cleaning house". This is necessary because the standing agreements, official (and unspoken), between local and state FOP/DAs and Police, are the greatest barrier to substantive and sustainable change with use of force policies and procedures. Breaking this barrier will require dismantling departments and starting over. This dismantling of police agencies, is not as a demonstration of power, but for administration reasons. It is necessary in order to free departments from those long standing deleterious agreements and relationships which promote cover-ups and lack of effective discipline. Letting go of all rank and file and leadership initially, will at least for a time, untether the cities and state, from those agreements, while new systems are put in place. Staff will be let go with pay and then required to reapply based on rigorous, new cultural competency standards. This systematic re-application process would require much greater screening and hiring standards reflective of the **core values of respect for the dignity of All life**. These processes would be designed to prize the majority of police officers and administrators who are demonstrating the highest

standards, and to weed out those who aren't. **This would be a relief for the vast majority, who would feel supported, and valued.** Increased funding will likely be necessary to ensure hiring and on-going training of outstanding officers.

2. <u>**Institutionalized, independent oversight bodies**</u> at the local level with investigative authority and a professional audit function are necessary to root out wrongdoing and ensure follow-through on discipline. Investigating individual police misconduct and auditing best practices at command level. Internal review i.e, internal affairs, is not optimal for objective review.... and for the most part....often due to conflict of interest, has not worked. Civilian review will better ensure objective evaluations, especially if the review boards are not beholding to any outside pressures. The selection of these individuals will need the utmost scrutiny to ensure objectivity, and will need to be continually evaluated as to effectiveness. This organization would hopefully, in time, be embraced by rank and file because it supports effective policing, rewards good behavior, working to promote good officers, weeding out the few bad ones, and improving daily operations.

3. **<u>**Training of recruits and ongoing education based on curriculum which explains how the history of systemic racism effects attitudes and behaviors.</u>** A

thorough knowledge of the culture of white supremacy values which founded this country and still runs it, in conjunction with the ongoing, difficult, emotional work required of white people to continually uncover our biases caused by those values, is necessary to optimally prevent violent, racist behavior. Curriculum must effectively address <u>accurate history of white supremacy culture</u>, which leads to hatred and fear of Black men, (i.e., Declaration of Independence excluding African Americans, indentured servants and women, and referring to Native People as "merciless Indian savages", Constitution counting Black people as 3/5's human, and Native People and women excluded, 12 Presidents owning slaves, connection of Fugitive Slave Acts to beginning of police force to arrest, imprison, and return runaway slaves to their owners / fear of Black men "raping our women", and under the Emancipation Proclamation, Lincoln wanting slaves returned to their owners in states loyal to the union) (2.) These curricula and training processes should be developed and overseen by a diverse group of experts in police reform - police, civilians, researchers, educators, trainers.

<u>Cultural competency requirement in recruiting and hiring.</u> The highest standards of cultural competency knowledge and skills is essential for carefully scrutinizing, recruiting and hiring of each potential police officer and administrator. **The challenges and responsibilities necessary for individuals who hold others' lives in their hands, must require the most rigorous cultural competency knowledge and skills along with being extremely emotionally balanced and mature.**

Hiring should minimally reflect the diversity of the community being served.

Rigorous Screening and Anti-bias training must include:
a. Police training research indicating shooting of blacks much more quickly than shooting whites.
b. Training in how to support each other in uncovering and addressing biases (i.e., not enabling, ignoring, or covering up, and broaching when hear, see, racist attitudes, words or actions).

However, implicit-bias training is not sufficient to mitigate against lack of accurate education about how white supremacy culture yields systemic racism in the best trained departments, (Minneapolis Police Department. apparently had high marks on their training program and polices). If we don't understand how we've been brainwashed to fear Black men, because they were seen as "subhuman, animals, who would "rape our women", automatic reactions to shoot first, and/or not be concerned about killing or doing serious bodily harm will continue.

4. Teams of mental health and health care workers and police would alleviate a heavy burden on police who are insufficiently trained to deal with some very challenging situations, like domestic violence, and severe health issues.

5. "Required" (not "recommended ") intervening and reporting of misconduct. This is essential for trust from the community, as well as for internal respect and trust. The blue code of secrecy is an insult to the majority of police who are individuals of courage, humility, integrity, compassion. But it takes much courage to stand up to others, especially if there is not the support of fellow officers, administration, or FOP. All rank and file and administrators need to sign written pledges/agreements which state, "I will not lie for you, I don't want you lying for me". Any officers or administrators caught lying should face termination. There cannot be self respect and trust within the rank and file, and certainly not with community, without this strict code of ethics. Enforcement of the highest codes of ethics is essential for each individual to proudly wear the badge which symbolizes the utmost responsibility to care for each other and each civilian with courage, integrity, humility, kindness and compassion. Transparency of all complaints, investigative process and disciplinary decisions should be made available to the public. Quotas for petty arrests and tickets should be abolished when evaluating professional performances and determining promotions.

6. Yearly "use of force" training. These are not one time skills. They need to continually be refreshed and improved on. Training must include on-going practice with the most challenging situations. **Testing of competencies would utilize carefully developed virtual reactions to realistic situations, especially in interacting with individuals and groups of different racial backgrounds.** The training should uphold highest standards, and require passing rigorous testing.

7. Licensing of all police. This would include highest standards of multicultural knowledge and regularly scheduled skill training, along with use of force training. It would also ensure records of use of force misconduct and terminations, to prevent jumping to other departments).

8. Regularly scheduled preventive PTSD, and other mental and emotional health training must be provided. Training would consist of how to deal with the contrast of everyday boredom / high adrenaline, stressful situations, how this is a set up for violent behavior on their parts, and how to best care for themselves in the light of the aforementioned, and in general. **Self-care has to be seriously valued by all officers and supervisors.** How this is reinforced is key. Of course modeling from the top is always preferable, if not essential.

9. <u>Training on toxic masculinity</u> for all level police. Emotional understanding about how the old macho norms for manhood affect the emotional and physical health of police, and contribute to violence is essential.

10. <u>**Truth and Transformation/Conciliation Process.**</u> **For hearts to change, for healing and trust to happen, on-going truth and conciliation process is necessary. The process should include acknowledgment of past misconduct/ use of power, sincere amends, realistic commitment to change, specific action steps to make it right, to ensure ongoing respect, safety, consequences for misconduct.**

**Policy changes will be temporary fixes without extensive screening and ongoing training which dives deeply into the systemic white supremacy values that founded the country and still run it, and live inside each of us. We must address the root / foundational causes of racism which live in our bones, DNA, emotions, psyches...often unconsciously. That is why it is so dangerous for people with weapons in stressful

situations who may spontaneously tap into uncovered, unhealed rage, fear....and pull the trigger or kick a dead person.

Borrowing from the acronym of the National Organization of Black Law Enforcement Executives ....policing is a "N.O.B.L.E." profession which requires those who live by the most basic, yet exemplary values of utmost respect for sacredness of All life. Adhering to the above recommendations will best ensure that our Peace Officers will be held in the highest positive regard by All of the people they "serve and protect".

<u>Prayer/ Wish For Our Peace Force</u>:
May each Police Officer feel supported in growing a strong sense of Love and compassion, in caring for her/himself and All other individuals..

Much appreciation and admiration for Dr. Timothy Knight, PhD (retired police officer, administrator, and consultant for police departments and community), for his wise counsel and support, in reminding us of how difficult the challenges are in working toward police reform, yet with courage, compassion and integrity, real change is possible.

Truth and Conciliation

Truth and Conciliation
can pave the way,
to heal the wounds
to a brighter day.

When we acknowledge the truth
a good place to start.
to expose the lies
that still tear us apart.

Sincere amends
from hearts that care,
to make things right
finally make things fair.

But the terrible losses
from generations til now,
all the rage and tears
being human will allow

And vetted reparations
no price too steep,
all the pain and suffering
the wounds so deep.

For trust to begin
we look each other in the eye,

and humbly, with hope
that makes us want to cry.

Sure beats the past
and continual sinning.
Is this the final answer?
No..........but.. a beginning.

Hell No.....but......a beginning.

P.S. The truth and conciliation/transformation process has been used since the beginning by individuals, families, nations..to heal and mend conflicts.... powerful, to be sure, if done with sincerity, and vetted by those who have been hurt and oppressed). Truth and conciliation ….."re"-conciliation is not really applicable here, since there was never togetherness from the beginning. Thank you for the reminder, Dr. Ruha Benjamin, Prof. African Studies, Princeton University, one of my heroes.

The following working paper was used as a guide to help in the preliminary stages of envisioning a truth and conciliation process in Princeton, New Jersey).

Preliminary Draft............Preliminary Draft................
Preliminary Draft............Preliminary Draft...

Truth and Conciliation* Princeton

This was a working document to be used as we embarked on our racial justice efforts. Some of the recommendations have been accomplished, some will continue to be addressed. *There cannot be "re"conciliation, when there never were harmonious relationships in the first place. Thus the more accurate term, "conciliation" is used here.

Introduction and Vision Statement:

The privileges and benefits of attending Princeton University and living in Princeton and the United States are marred by persistent racism. However it is clear, that racism in the US is alive and well. There are also many strong benefits to all of us in acknowledging and addressing racism and white privilege. When everyone feels valued and respected, life is better for all.

Much of this racism stems from the refusal to acknowledge the accurate history of the founding of this country—viewing and treating enslaved Africans as property and sub-human savages, not people, and the genocide of the Indigenous people.

The equality of all men written in the Declaration of Independence masked the reality of where power and control resided. The Constitution, however, reveals the support for a ruling elite of white heterosexual male property owners. Missing from the core national values was respect and appreciation for the dignity of every human being and for all life. The question is are we ready and willing to confront racism in a genuine way by reclaiming that respect for all life as our current national core value? The truth and reconciliation efforts in places like Australia, Canada, and Africa have begun the process of healing from racism. Perhaps we in Princeton can also provide a model of reconciliation for other communities and for our nation.

At the heart of truth and reconciliation is a genuine wish to promote healing of past and current racial trauma wounds, and to ensure ongoing equity and compassion for **all** people. Healing racial trauma, as with any trauma healing, first requires the honest, often painful, acknowledgment of the actual atrocities which occurred and continue to occur. As James Baldwin said, "We can't change what we don't face." Along with the honest acknowledgment of the past and current abuse, a sincere and compassionate demonstration of action—"reparation"—which addresses an effort to promote equity and justice, is necessary for real forgiveness and trust. Ongoing oversight to ensure consistent follow-through of proposed actions is of paramount importance.

Holding ourselves and each other accountable for mending centuries of painful injustice is no easy task. It suggests a deep, humbling look at our personal and common humanity and imperfections. If we of the Princeton community and schools are willing to courageously and compassionately acknowledge the real history of racism and the denial of white privilege that prevents true healing, our lives and the lives of our children will be significantly improved.

Now is the perfect time to make our mark in addressing racism in Princeton. A powerful, cooperative partnership among the community, schools, and University is necessary to ensure success. Superintendent Cochrane, Mayor Lempert, and President Eisgruber have each expressed intention to support racial justice efforts. We call on each of you to provide superlative wise, courageous, and compassionate leadership in this most important effort.

We of Not In Our Town Princeton appeal to our fellow community members of the municipality of Princeton, Princeton public and private schools, and Princeton University to come together to eliminate racism. Now is the time. If not now, when. If not us, then who? As Margaret Meade said, "Never doubt that a small group of dedicated individuals can

change the world. Indeed, it's the only thing that ever has."

Given the aforementioned vision statement, the following is a list of core principles, values and strategies:

Background and Guiding Principles:

This is not about guilt or blame. It is about acknowledging the truth so we can finally begin the healing.

The US government, Princeton township and Princeton University were primarily established for the wealthy white male elite, not for African Americans and Indigenous peoples / First Americans.

The core values reflected and continue to reflect white privilege/advantages/benefits of power, wealth, control, entitlement, elitism, competition to be number one at all costs no matter who you need to walk over to get there.

The horrific treatment of enslaved African and Indigenous people served as a powerful example and warning to immigrants of the need to "assimilate", that is to adopt the "dominant" culture's norms, often at the expense of the traditions and values of their cultures. As with most colonization, this set in motion a pattern of the various immigrant groups competing against each other to avoid the oppression and discrimination of being on the bottom rung of the white supremacist constructed racial hierarchy (like Blacks and Native people). Erroneous racial categories were invented to better serve the white ruling class dominance. Since the government was established to serve the wealthy property owners, poor and middle class whites have continued to feel resentment, even hatred at not receiving their fair share of the wealth. This led to fear and mistrust of African Americans and Native people who were and still are perceived as the key threat to the white piece of the pie, rather than the wealthy power elite. This continues to be a clever manipulation indeed.

Truth and Conciliation Plan

We believe that a kind and just Truth and Conciliation Plan embodies three core elements:

1. **Truth** acknowledgment of the accurate history of racism in this country and in Princeton, and related consequences, the results of which are still being experienced.

2. **Conciliation** – Offering apology, sincere regret for the injustices, commitment to correct the wrongs

3. **Reparative Strategies** - demonstration of effective, consistent, timely action which reflects this commitment to promoting equity and justice, and whenever possible, the return of wealth or property that acknowledges the original crime.

Therefore:

We commit to ongoing acknowledgment and reporting of the accurate history of racism in in this country and here in Princeton, and related consequences, the results of which are still being experienced.

We acknowledge that the Truths mentioned below did occur and continue to effect the lives of everyone in this community, most especially African Americans and Hispanics. We humbly offer our heartfelt apology, sincere regrets, and intention to correct the wrongs.

We are committed to providing an ongoing demonstration of effective, consistent, timely action to reflect the aforementioned intention to correct the wrongs. This action will be driven by and approved by the residents of the Witherspoon-Jackson Neighborhood.

The following inter-related issues require our utmost attention and energy in addressing Truth and Conciliation, and Reparative Strategies*:

1. Slavery in Princeton

2. Education

3. Housing

4. Employment / Economic Disparity

5. Business

6. Government

7. Healthcare

8. Churches

9. Police

10. Everyday Micro/Macro-Aggressions

1. Slavery in Princeton

The horrific, economically motivated institution of slavery informed and continues to inform the treatment of African Americans in the areas of education, housing, economic and employment disparity, and the persistent level of everyday biases to which African Americans, are subjected to. The institution of slavery has been replaced with the billion dollar institution of mass incarceration of a whole generation of Black men and Native men, many of whom were also slaves, for mostly non-violent crimes. This separating of men from their families was a conscious effort weaken the Black family and culture. This imprisonment prevents them from getting jobs because they then have criminal records. And the cycle of slavery continues.

Slavery did exist in Princeton. Up until very recently, this truth has not been known or taught, certainly not in a concerted way. Some examples of the history of slavery in Princeton, and at Princeton University are that Presidents of Princeton University, township officials, and business heads, owned slaves. A 2008 senior thesis, which Princeton University Professor Martha Sandweiss assigned as a classroom text and considers well-documented, argues that Princeton's first eight presidents were slaveholders; one of them, John Witherspoon, shored up the University's post-Revolutionary War finances with money from slave interests. Slaves contributed to the building of Princeton

University. The slaves who worshiped in the balcony at the First Presbyterian Church (Nassau Presbyterian Church) started worshiping at First Presbyterian Church of Colour in Princeton (now Witherspoon Street Presbyterian Church, where distinguished NIOT member Shirley Satterfield, has worshiped since childhood). In 1837 Elizabeth "Betsey" Stockton, when she returned to Princeton, taught the colored children and started a sabbath school at Witherspoon.

Reparative Strategies For Slavery In Princeton

a. Commemorate the slaves of Princeton and their ancestors in yearly ceremonies, and other just reparation, such as financial compensation, housing, scholarship. Courts often give compensation for individuals and families of those who have been abused by authorities, wrongly imprisoned or killed.

b. A very important series created by Shirley Satterfield, of "From Slavery to the Grave in Princeton" for the Evergreen Forum at the Senior Resource Center should be attended by all, and hopefully archived for viewing, and along with many of these programs mentioned, serve as model programs. Professor Sandweiss of Princeton University is one of the presenters for this series.

c. Shirley Satterfield's walking tour of the African American history in Princeton, sponsored by the Princeton Historical Society, starts at the campus in front of Nassau Hall to tell about the early college presidents who owned slaves.

d. A very important course is being created at Princeton University by History Professor Martha A. Sandweiss about slavery in Princeton. Working with important individuals who grew up in Princeton, like Shirley Satterfield, who attended the segregated schools in Princeton, and then became a school counselor at Princeton High School, lends further credibility to this important work. As this course continues to grow, it will hopefully include other important individuals who grew up in Princeton.

e. Prominently displayed signage about slavery in Princeton and Princeton University. Identifying presidents who owned slaves, and contributions of African Americans to American and Princeton culture.

f. Ensure accurate history about slavery be taught at all levels. For example, the real reasons for Civil War, economic/freeing the slaves (slave labor undergirded the U.S economy.). As slaves were freed, the klu klux klan and other hate groups grew in number to "protect their rights". Thus the original manifest destiny doctrine, combined with fear of the "angry Black man" can be viewed as key components to developing implicit or unconscious bias towards African Americans as well as many other groups (such as Asians, Latinos, other immigrant groups, LGBTs, women, people with disabilities), which may be perceived as a threat to white supremacy. An effective colonization strategy used to control the masses is by pitting them against each other.

2. Education.

While not the sole responsibility of the educational system, quality education can be a powerful vehicle for increasing racial awareness, sensitivity, and healing for the entire community.

-Segregated schools existed in Princeton until 1970? Many Native children were ripped apart from their families and villages, forced to attend Indian schools (like the Carlisle Indian School),where they had to cut their long hair)a symbol of native pride), give up their traditions, dress in constrictive uniforms, and severely punished (or even killed), if they spoke their native language or tried to escape.

-The quality of education for African American students since integration has been questionable, ie.. students often automatically placed in special education / "mastery" classes (another name for "slow" or "challenged" learner), and discouraged from considering college.

-There is a significant "achievement" gap for African American and Hispanic students.

-Racial bullying and micro/macro aggressions, continue to exist from both other students and teachers.

-Princeton University is built on the homeland formerly inhabited by the Lenni Lenape Indians, (and likely not given up voluntarily). The Lenni Lenape consider themselves "the First People".

-Students at the Woodrow Wilson School of Public Policy and International Affairs are continually insulted and disrespected by attending a school named after a segregationist and ally of the KKK.

Reparative Strategies For Education-

a. Provide Racial Literacy courses and training for all citizens of Princeton, including all students in the public and private schools and at Princeton University, parents, township officials, Board of Education, university Board of Trustees, public school and university administrators, faculty, staff. (Princeton University Professor Ruha Benjamin).

b. Provide training in unconscious/implicit bias, microaggressions/ macroaggressions, how to have difficult discussions about race for all teachers, administration, staff, parents, students. (Howard Stevenson, Beyond Diversity Resource Center).

c. Establish "safe havens" and peer mentoring groups for students and staff, to support victims of racial bullying, the bullies, and parents of the bullies. Specialized training and careful selection process is required for these mentors.

d. Hire African American counselor and more African American teachers at all schools.

e. Establish procedures and consequences for tracking incidents of racial bullying and discipline cases by race.

f. Establish school blog where difficult discussions about race and strategies can be aired.

g. Create prominently displayed signage stating that:

    7. African American and Native American students attended segregated school at this cite.

    8. Princeton public schools remained segregated, and African Americans and Native Americans (and women), were not allowed to attend Princeton University, until the middle of the 20th century.

    9. Princeton University is built on land formerly inhabited by the Lenni Lenape Native Americans.

    10. The Woodrow Wilson School of Public Policy and International Affairs is named after former president Woodrow Wilson who was a staunch segregationist who was aligned with the KKK.

h. Provide strong funding to support Shirley Satterfield's walking tours of the African American history of Princeton, tour on tape, training of others, archives as part of Historical Society, and training of Princeton community and infused into the k-12 curriculum and Princeton University core curriculum.

Include Princeton University Professor Kathryn Watterson's important book on the history of the Witherspoon-Jackson 20th Historical District, "I Hear My People Singing: Voices of African American Princeton".

i. Provide college scholarships for deserving African American and Hispanic students, especially for descendants of slaves and Native people, as other universities, like Georgetown are offering.

j. Promote awareness and public education of all citizens about the white privilege value system and its impacts, through educational series at public library, school courses, films at Garden Theatre.

k. Support commemoration of former students and their families through programs like the Unity Awards and others.

l. Continue other important initiatives Superintendent Cochrane has supported and promoted such as Day of Dialogue with teams of students and staff from each highschool in Mercer County, developing a Diversity Institute for Mercer County, and infusion of the CHOOSE racial story telling curriculum into all the Princeton public schools.

m. Produce a joint report including recommendations to the Princeton Board of Education, concerning the history and experience of living and going to school in Princeton, experiencing structural racism and white privilege, (including systemic harms, inter-generational consequences and the impact on human dignity) and the ongoing legacy of the Black experience in Princeton.

n. Emphasize Paul Robeson's biography, performances, and writings in the Princeton schools, especially in Princeton and U.S. history classes.

o. Examine core values of Princeton public, private and higher education i.e., providing accurate history/racial literacy, honoring all students, building self-esteem, decision making skills, relational skills, asking each student to do racial justice service project.

p. Continue NIOT educational related activities such as the monthly Continuing Conversations series on Race and Privilege with the Princeton Public Library, yearly Unity Awards given to students who demonstrate outstanding projects and examples of addressing racism and privilege, monthly meetings with the Superintendent to discuss such issues as racial literacy training, addressing racial bullying and minority hiring practices in the school district, participating on the Joint Youth Services Committee, collaborating with the Princeton Civil Rights Commission, and offering consultation and support to other communities.

3. Housing

Historically and currently there continues to be a significant lack of desirable and affordable housing for African Americans and Hispanics in Princeton. There has been a concerted effort to eliminate the historically African American Witherspoon-Jackson community. Due to herculean efforts of Shirley Satterfield, President of the Witherspoon-Jackson Historical and Cultural Society, this community was finally designated as a historical district. However this does not make up for the many individuals and families who were forced to leave Princeton due to an inability to afford the continually increases in taxes, rents etc. This district continues to struggle to maintain it's dignity and adequate standard of living.

Reparative Strategies for Housing-

a. Acknowledge the history of intentional segregated housing. As Native people were forced to live in deplorable conditions on the reservations, African Americans have been economically and socially forced to live in segregated communities.

a. Provide upgrades and financial support for those still living there.

b. Increase permanent affordable housing availability for those with lower income levels, rather than continuing to provide "affordable housing" for those with higher income levels.

c. Provide housing for those forced to leave due to raised taxes and rents.

d. Investigate feasibility of having the municipality issue a moratorium to return the tax level for "grandfathered" African-American residents to the 2010 effective rate of 1.732 and have it maintained for the next 20 years as long as the descendants remain primary owners and inhabitants of the properties. (2016's effective rate is 2.577.)

4. Employment / Economic Disparity

There has never been equity in Princeton regarding employment opportunities and economic advancement for the African American and Hispanic community.

Reparative Strategies for Employment and Economic Disparity-

a. Provide job incentives, preparation and training for those in need.

b. Provide increased job opportunities in the community, in the public schools and at the University.

c. Give first consideration to members of the African American and Hispanic community for job openings.

d. Provide financial planning, and bankruptcy coaching for those in need.

5. Business

There has been a significant history of discrimination by businesses towards people of color, especially African Americans. For example after returning from military service being turned away at restaurants, and discriminatory hiring practices and treatment of people of color.

Reparative Strategies for Business

a. Participation in yearly T & C ceremonies, by business owners associations and all business owners.

b. Provide RL/IB training for all business owners and staff.

c. Ensure equitable hiring practices for all businesses.

6. Government

There has been a significant history of discrimination by the local government towards African American and Hispanic communities.

a. Acknowledgement of history of discrimination

b. Participation in yearly T & C ceremonies..

c. Ensure equitable hiring practices and representation in government.

7. Healthcare

Healthcare disparities with the African American and Hispanic community need to be addressed immediately.

Reparative Strategies

a. Acknowledge the history of healthcare disparities, with specific improvement plans and timelines.

b. Participation in yearly ceremonies by healthcare agencies

8. Churches

Churches vary in their historical commitment to addressing racism.

Reparative Strategies

a. Acknowledge accurate history of each church in addressing racism.

b. Participation in yearly T & C ceremonies by all churches.

c. Address lack of minority membership in respective churches.

9. Police

There has been a mixed relationship between police and the African American and Hispanic communities.

Reparative Strategies

a. Acknowledge history of treatment of African American and Hispanic communities.

b. Ensure adequate representation of African American and Hispanics on police force.

c. Continually monitor racial profiling.

d. Continue to emphasize community policing initiatives.

10. Everyday micro/macro aggressions-

Members of the African American and Hispanic community have always been subjected to many often subtle, unintentional yet nonetheless hurtful and demeaning everyday comments, gestures, behaviors, or microaggressions, with little or no consequences to those who inflict them. This is a common experience for many on the streets, in stores, in the schools, in interactions with police, and in dealing with township bureaucracy. Sometimes these offensive experiences rise to intentionally hurtful or macroaggressive acts.

We must all do a much better job of holding ourselves and each other accountable for understanding where this behavior stems from, that is learning about our implicit or unconscious bias, and commit to learning how to reduce and eliminate it.

Reparative Strategies for Everyday Micro and Macro Aggressions-

a. Provide community-wide training for all the above mentioned groups in implicit bias, micro and macro aggressions, non-violent communication, and how to have difficult discussions about race in a respectful and compassionate way.

11. Safe Haven for Immigrants

Princeton has stepped up to be a safe haven municipality. This is an extremely important initiative that needs to continue.

*An inaugural commemorative ceremony (like the recent one at Georgetown University), would be a wonderful way to initiate the truth and conciliation effort. It would demonstrate in a very public way the commitment to recreate a community, school district and university which feel more inclusive, and which we **all** can be proud of. Yearly ceremonies would accentuate and reinforce community solidarity.*

- For the above to be an effective, vital, living, growing initiative, regular oversight by a group composed of representatives of all the aforementioned constituencies, is absolutely necessary. Each constituency will provide at least two members to be part of this on-going working group.

Final Remarks and Hope For the Future
None of the above are easy or quick cures. However with a genuine, openhearted willingness to look at ourselves and each other with compassion and kindness, we will be able to make great strides over time. Integrity, patience, understanding, courage, gentleness, mutual support, accountability, encouragement and faith in the goodness and good will of us all, will definitely help to begin healing the traumatic wounds of racism, so that our children and their children will have a happier, safer and healthier Princeton. Please join us in this loving endeavor. Thank you.

We the undersigned agree, on this day, to begin to undertake the above mentioned Truth and Conciliation Plan, to the best of our abilities. This is our word.

~

Trust

Trust,
like forgiveness,
is a gift,
I choose
to give myself,
or another...
when I'm clear
of my commitment
to myself.

That I'm always there
for me.....
dependable.
or the other
is there for me.....
dependable....
with my best interest
at heart.

It's not a perfect equation.
Because sometimes
we fall short.

But when I do,
I hold myself
accountable
with love,
and clean it up.
And ask the same
of the other.

It's an agreement
with myself,
with the other.

Yes.
That's it!.

An agreement
to Be
Dependable.

Thank you.
You're welcome.

3/2020

The Sacred
Back to the Beginning

<u>All</u> life is Sacred
Wise ones have always taught
This simple, beautiful truth
Can't be sold or can't be bought.

From birth to death
Our oath, life giving pledge
To treat All life as Sacred
Re-Creation from the edge.

We've felt alone and lost
However, not <u>too</u> far,
With <u>Love</u> our guide forever,
We Re-member....who....we.....are.

So, awakening from the sleep,
We greet this brand new day.
And the journey begins anew...
Inside me
And inside you.

Inside me
And inside you.

Mitakuye Oyasin
All My Relatives
May we Re-member.

On the heels of the last poem "**The Sacred**", I felt moved to add this proposal, with hope that all those in power (including you and me), will consider going back to the beginning, to boldly, lovingly, recreate a new vision for ourselves and each other. A Loving Vision of the Sacred... for Healing America...and our world. I offer you this, with hope and faith in our individual and collective commitment to living our lives with full love....Being all we truly can be.

This proposal is offered, with hope that all those in power (including you and me), will consider going back to the beginning, to boldly, lovingly, recreate a new vision for ourselves and each other. A Loving Vision of the Sacred... for Healing America...and our world. I offer you this, with hope and faith in our individual and collective commitment to living our lives with full love.... Being all we truly can be.

ALL LIFE IS SACRED

A Loving Blueprint
For Healing America

"I am poor and naked, but I am the chief of the nation. We do not want riches, but we do want to teach our children well. Riches would do us no good. We could not take them with us to the other world. We do not want riches. We want peace and love"
-Red Cloud (Makhipiya)
(late 19 century) Lakota Chief

Introduction: The Original Core Values

To Review:.......While our Constitution espoused positive sounding values of liberty and justice for "All The People"... Women and Native People were excluded from the document, and African American males were counted as 3/5th of a person. Beautiful values and guiding principles of Love - honesty, fairness, kindness, compassion, generosity, respect.... were largely forgotten in the desperate quest for power, wealth and domination.

So in actuality, at it's core, the U.S.'s founding values were more indicative of fear, greed, hatred, material wealth, power and control, which further devolved to elitism, entitlement, superiority. "Success", was equated to competing to be #1, at all costs, no matter who we need to walk over to get there. Those initial values were derived by the misguided principles of "doctrine of discovery" by "god's anointed ones", justifying "manifest destiny". Those principles resulted in the genocide of the Indigenous People, the original inhabitants and stewards of the land, and the horrific inter-generational treatment of enslaved Africans upon whose backs our country's economy was largely built. Additionally, the careless slaughter of the animals and desecration of the land....all for the use of those in power.

The aforementioned paved the way for protecting those in power by establishing and justifying institutional core values of racism, sexism, homophobia, classism, xenophobia, ableism, ageism and on and on. So yes, we have many blessings, and of course, we also have many severe problems. So that's the quick and very dirty, explanation for the state of affairs we find ourselves in. Now.....what do we do about it?

Choosing New Core Values To Live By:

All Life Is Sacred

Again, as I see it...the answer is really quite simple....and yet.... profoundly challenging. We each choose to claim, and live by, the incredibly powerful, core, loving principle......."All Life Is Sacred"....Sacred...meaning worthwhile, important, valuable, precious, worthy of love, respect, dignity, awe and wonder. This also translates into this new perspective that I'm no better, no worse than that insect, the Land, that tree, that Woman, that child.....**ALL Life Is Sacred.** Once we live by the love principle, all of the poisonous "isms" would die out. The sacred interconnectedness of all life...takes center stage..... we literally live by the idea of **"All interconnected, All One"**. This has always been a core spiritual principle. Now, the science of quantum mechanics demonstrates that at a cellular and molecular level, all of our cells and molecules are always changing and intersecting with all others. So we literally are, all connected and all one.

From the Lakota Native Spiritual Tradition...Mitakuye Oyasin (pronounced – mee tock o yay – o yah sin), meaning "All My Relatives, All My Relations, "We Are All Related...All One"... Each One my Sister, Each One my Brother....The Animals Are My Relatives...The Trees....All part of the Sacred Circle Of Life, The Sacred Hoop...Mother Earth, Father Sky, Grandfather Rocks, Wind Spirits, Thunder Beings....all Guides and Teachers from the Great Beyond.....All To Be Honored and Cherished...and All to live with in Harmony. So I do my best to Be Grateful for the Sacred Gift Of Life, and for all the Gifts...My Relatives....and choose to release differences and judgments of myself and all others.

> *"Lose all differentiation between myself and others, fit to serve others I will be. And when in serving others I win success, then will I meet the Buddha..and we will smile"*
> *-Milarepa, The Great Yogi of Tibet*

> *"You will treat the alien who resides with you no differently than the natives born among you; you shall love the alien as yourself; for you too were once aliens in the land".*
> *- Leviticus 19:33-44*

So if this was the core guiding principle which drives the purpose, the vision and mission statements of every school and church, and all the other institutions in our country... and the world...imagine... please imagine....the affect and effect on all of our children, and their children's children for the next seven generations to come...... We will treat ourselves and each other with utmost gentleness, kindness honesty, respect, fairness. Because we are all in this together...we do not see anyone as an adversary in competition for a little piece of the pie. There is plenty to go around for everyone to have a high quality of life, when we choose to live with utmost mutual regard for the sacredness of all beings.

This is a system of values which is in **everyone's** best interest... individually and collectively. Of course, this would require a full-on **agreement**....and full-on **commitment...** by everyone. A **100% commitment** to being our best, with complete honesty and compassion in every thought, word and action, in how we treat ourselves and each other. A high bar to live by ....to be sure. And yet.....we would have soooo much to gain....every one of us. And living by this guiding principle of the sacredness of all life creates this incredible opportunity for "**Loving** Thy Neighbor **As Thyself**". That's it. As simple as I can put it.

So now, a little more about this thing called **Love**.

Love

Since the beginning, all the wise ones have said …..Love **Is** the answer.

So, what is this elusive butterfly called Love?

As the inimitable Tina Turner sang..."*What's Love Got To Do With It?*".......the answer......Everything!
The most powerful force in the universe is LOVE! Always has been, always will be. We humans just forget. And to be fair, it's mostly because we haven't really been taught about love......

not nearly as we need to be. The word is thrown around so carelessly, most often without really even knowing what we mean by it. *"It's just a feeling....can't really be explained....but you know it when you feel it"*......like that. Talk about bogus. We've been brainwashed, bamboozled, hoodwinked......to desperately seek this ....thing......that we can't even describe...... and we wonder why so many relationships with ourselves and each other end up not feeling fulfilled. As the wise and courageous activist and writer bell hooks (lower case, her preference) so heart-fully teaches us in her wonderful book "All About Love".....we need to define the core values of love (my words, not hers), and then we can create lives filled with those values. Values like gentleness, kindness, commitment, integrity, courage, vulnerability, dependability, fairness, compassion, understanding, responsibility, gratitude, trust, and being willing to support ourselves and each other in our mutual spiritual growth. Spiritual growth, for me, implies a life-long learning adventure in deepening my inter-connectedness with all other life, so that we may best serve all others.

So we're not talking about mamby pamby, goo-goo, ga-ga fantasy love, we're talking about the real deal....fierce, active, passionate, clear, on solid ground LOVE. And it has to start with us being taught about how to honor, and cherish, respect, and care for **ourselves**. And holding ourselves to the highest and most honorable standards of human decency and fairness. If we are willing to bravely venture into this adventure in learning how to love ourselves....and to share that with others as we grow......which, to me, is the main purpose for our being born,.....then all good things are possible.

So there you have it sports fans! That, to me, is the essence to our living harmoniously with ourselves and each other, on our Sacred Mother Earth. The question then becomes...if we agree, in principle about the aforementioned core values of the sacredness of all life, and of living in love as the blueprint for honoring the sacred....then what are we willing to do about it?

For me it's all about the powerful word......"COMMITMENT". Am I....Are You....Are We....each individually...and collectively......ready.....willing.....and able to commit to living our lives in beauty...walking the Sacred path of Love....in all of our thoughts, words and actions............ ...individually and collectively? If we are....we can fix the problems we've co-created.....if we are not.....then we will continue to have the unfulfilling existence we have. Again, commitment, to me, means....100%...all in... not 99%....for that 1% will sabotage our best efforts. That doesn't mean we will do it all perfectly... that's not the point. It's that when we mess up, which we will....no big drama, no excuses.....just, pick ourselves up, dust ourselves off, look honestly and humbly at what we missed that created the misstep, acknowledge it, come up with a better strategy for fixing it expediently, and then carry through with a sustainable plan. In Twelve Step Programs we call it taking a fearless and searching moral inventory and making amends, and committing to and following through on real change. This is similar to the Truth and Conciliation process which so many individuals, families, communities, nations have used to great avail. (Please refer back to some of my prior poems and pieces on truth and conciliation if you're interested in reviewing a little bit more about these and other similar, complementary processes).

I don't know about you...but I'm ready, willing and able to commit to co-creating the amazing life here, and around the world, that I believe we were all put here to live. I hope and pray, that you are too....and that is why you are still reading this.

Real love is humble, and gentle and kind, courageous, and impeccably honest. The following are guides to live by, to create inner well-being and outer well-being for us all. So let's look at some of the core ingredients of love and what they represent, (at least to me). And as we explore these core values more and more, it becomes clearer how they each intersect with and complement each other:

"Out of the Indian approach to life there came a great freedom – an intense and absorbing love for nature; a respect for life; enriching faith in a Supreme Power; and principles of truth, honesty, generosity, equity, and brotherhood as a guide to relations".
Luther Standing Bear (1868-1939)

Oglala Lakota chief

So let's look at some of the core ingredients of love and what they represent, (at least to me)
And as we explore these core values more and more, it becomes clearer how they each intersect with and complement each other:

Core Love Values

Courage- The willingness to lean into the difficult challenges of life, to Be, and to Do the "right thing". Even in the midst of others' disapproval. A great guiding question for any quandary in life is...."What would love do now?" "What would be the most loving thing I can do for myself in this situation....that little still voice of truth inside, will always guide us home.....to love....if it doesn't feel right, then the guide isn't love, it's usually fear or guilt. Really important to keep re-learning about that distinction. Are we willing to be courageous enough to be vulnerable...with our feelings....this is the mark of a true love warrior.

Honesty- A strong promise to ourselves to act and speak with authenticity. No white lies. When we "get over" on others we're really getting over on ourselves. No way we can feel good about ourselves without living honestly. "The truth **will** set us free". Which is why it is so essential to live by this core value. Lest we be imprisoned by our secrets. "We are as sick as our secrets" (from 12 Step Principles).

Integrity- Living with a strong sense of justice and fairness. We live congruently....our behaviors match our values. And again, a willingness to speak up and take action when we see or feel inequity and injustice. Living with impeccable honesty and integrity frees us to live in peace with ourselves, and to feel a deep sense of respect and acceptance.

Kindness - We treat ourselves, and others with gentleness, compassion, generosity, empathy, understanding, support. Again, it is crucial that our relationship with ourselves is based on utmost kindness. Otherwise our treatment of others, will likely fall short, and ring hollow, eventually, even leading to resentment and sabotage. Are thoughts and actions reflect our desire for the well-being of ourselves and all others.

Commitment - We consistently follow through on our promises to ourselves and others, especially when the going gets tough. And with all these, when we fall short, no big drama, no beating ourselves up, compassionately acknowledging where we went wrong, and revising strategies that are perhaps more realistic and more sustainable.

Dependability - We do what we say, and say what we do. Similar to integrity. We believe and have faith in ourselves, because we carry through on our commitments...first and foremost with ourselves, and of course also with others.

Responsibility - We take pride in being accountable to ourselves, first and foremost...honoring our most important love priorities in self-care, and the care of others. Being responsible for our well-being fills us with a sense of contentment and peace, since we know we are "taking care of business". Having clear boundaries is very important. If we don't know what our limits are, and don't honor them, we can become resentful, and being doing things out of guilt and fear, not love.

Fairness/Equity/Equality – We treat ourselves and others with a sense of fairness, equity and justice. We do our best to ensure equal and fair access to life necessities for all, economically.. food, clothing, shelter, educationally, healthcare...a desirable quality of life for all. If one of us is suffering, it hurts our common humanity.

Again, there is plenty to go around if we live by fair and just standards.

Patience and Acceptance – We do our best in all of our thoughts words and actions to be very patient with our imperfections, and to hold ourselves and others with unconditional positive regard. We do our best to check our judgments of ourselves, to treat ourselves fairly. And we strive to do the same with others, by trying to put ourselves in their shoes. While this can be extremely difficult....like with all of these values, it can also be extremely rewarding, as it gives the best chance of interconnecting with others who may seem different than us.

Trust - Trust takes time to build for ourselves and others. When we consistently demonstrate over time, that we are trustworthy, that is true to our word, then trust becomes a great gift we give ourselves and others. When trust is broken, this is how we determine our willingness to recommit to our core principles. We can do this by carefully accessing what let to the break in trust, and what needs to happen to regain the trust for ourselves and the other. A sincere, heartfelt amends, when appropriate can be very helpful. And then following through on well-thought out strategies to make things right for ourselves and with others (ideally, vetted by the injured party). And of course, consistently carrying through on our commitments to change, and monitored over time. If need be, reassessing and altering strategies as needed.

Respect – Treating ourselves and others in the way we would like to be treated. Being respectful and considerate of our and others' feelings, and needs, as consistent with our core values of love and sacredness of all life. Treating ourselves and each other with dignity reverence for each one's worth. This can be especially difficult when there is significant disagreement, Yet, most disagreements can come to fair and equitable resolution, if both parties are committed to treating each other with respect and kindness, and a willingness to meet in the place of equity and fairness.

Gratitude/Affection – Wise elders often say that gratitude is one of the most important love values there is, and can be a true source of genuine affection which we all so greatly need. Affirming and acknowledging our self care, along with our gratitude to the other, can be beautiful and heart filling gifts which make a huge impact on our mental, emotional, spiritual, physical and social well-being.

And relationships individually and collectively can prosper and grow through this simple act of loving expression.

Spiritual Growth - Willingness to nurture spiritual growth in self and others - When we choose to believe in and live by a belief in the core worth of ourselves and others, and of our interconnectedness to all things, then our life becomes a reflection of that core value, by how we treat ourselves and each other in our daily lives. Decision making about the most simple, and the most difficult challenges becomes more achievable as our confidence grows through the experience of our spiritual growth.

Core Love Values Inventories

Using the Core Love Values Listed Above, please consider filling out the following two Core Love Values Inventories For Self and Others, and for Leaders, Institutions and Organizations. For the second inventory, you might start out by picking one leader, institution or organization that moves you to want to take action, and see what comes up for you. (Institutions and organizations may include any groups such as schools (especially teaching accurate history), healthcare, wildlife, environment, churches, business, police/prisons (of course would include treatment of inmates), military, government (especially promoting truth and conciliation processes), and others. In service of all the people, all leaders, institutions and organizations would optimally be reviewed internally and externally (perhaps by unbiased, carefully selected, civilian review boards) for realistic compliance and sustainability on an annual or bi-annual basis. And,

Men and Racism: The Healing Path | 173

for example, a mailer could be sent to cross sections of the state populations or townships asking for feedback using the inventories for whichever leaders institutions/organizations are up for review. Feedback could then be summarized and used for regular review and planning sessions. Citizen ownership would then be a vital, ongoing, mutually advantageous process. Data from both inventories help us determine what is working, and improving what needs improving. This is not about being adversarial with ourselves or others, it's about working together for optimal outcomes for all.

All Citizen's Core Love Values Inventory For Self and Others*

Core Value	Rate how treat self/others from 1-10 (10 being optimal)		Describe recent time demonstrated that value		Describe specific plan to improve that value	
	Self	Others	Self	Others	Self	Others
Courage						
Honesty						
Integrity						
Kindness						
Commitment						
Dependability						
Responsibility						
Fairness/Equity Equality						
Patience/Acceptance						
Trust						
Respect						
Gratitude/Affection						
Cooperation/Service/Sharing						
Spiritual Growth						

*Based on Fearless and Searching, Kind Moral Inventory from Schiraldi, G. R. (2011), *The Complete Guide to Resilience: Why It Matters; How to Build and Maintain It.* Ashburn, VA: Resilience Training International. © 2011 Glenn R. Schiraldi, Ph.D. Not to be reproduced without written permission** *Human Options.* Toronto: George J. McLeod Limited, 1981, p. 45.

All Citizens' Core Love Values Inventory For Leaders, Institutions, Organizations*

Core Value	Rate how leader treats employees and public from 1-10 (10 being optimal)		Rate how Inst./Org. treats employe/pub. (from 1-10)		Describe recent time demonstrated that value		Describe specific plan to improve that value	
	Emp	Pub	Emp	Pub	Leader	Inst./Org.	Lead.	Inst./Org.
Courage								
Honesty								
Integrity								
Kindness								
Commitment								
Dependability								
Responsibility								
Fairness/Equity Equality								
Patience/ Acceptance								
Trust								
Respect								
Gratitude/ Affection								
Cooperation/ Serv./Sharing								
Spirit.Growth								

*Based on Fearless and Searching, Kind Moral Inventory from Schiraldi, G. R. (2011), *The Complete Guide to Resilience: Why It Matters; How to Build and Maintain It*. Ashburn, VA: Resilience Training International. © 2011 Glenn R. Schiraldi, Ph.D. Not to be reproduced without written permission** *Human Options*. Toronto: George J. McLeod Limited, 1981, p. 45.

Tree Of Living Love

The following is a visual representation of the Tree Of Living Love – a symbol of the mutual commitment to live and nurture ourselves and each other... All life. One way to view this is to envision a truth and conciliation process, that is, speaking the truth about our history, and seeing it all, through the eyes of a gardener and the garden of life. We live in a beautiful garden. There are many gorgeous flowers and plants of all colors and shapes. There are also invasive and aggressive weeds that may look attractive, however, left unattended, may drain the soil of it's nutrients, and overpower and suffocate the natural healthy and balanced growth of all of the flowers and plants. It may be noted, that the poison of some of the weeds, blended with the sweet pollen of some of the flowers and plants, can have powerful healing properties. So the gardeners need to first learn about and recognize the poisonous roots of decay. And then the garden needs continuous, vigilant weeding and tilling and revitalizing of the soil to maintain the precious life affirming balance.

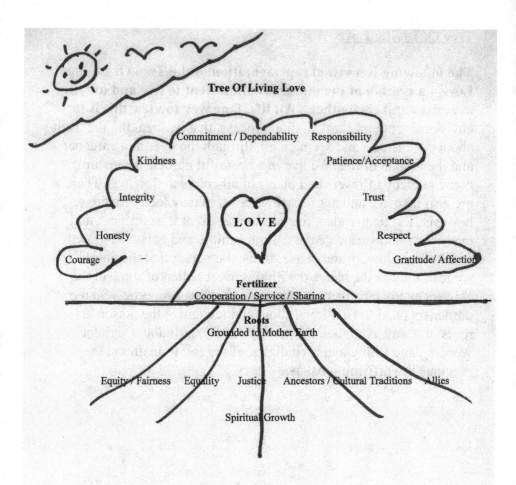

Commitment Proposal

It is proposed, that every citizen will sign and commit to live by the following agreement(whether born here or not... refugee, immigrant, young, old, gay, straight, parent, child, blue color worker, white color worker, educator, government servant, police/peace keeper, business person,...and on and on)....in other words, everyone of us.

All Citizens' Agreement*

I, as a citizen of this Country, of this World, as a representative of ...All The People....Pledge my life, to do my best with every thought, word and action... to make choices based on the core belief that All Life Is Sacred, and that each choice will reflect core Love values by:

1. Checking and accessing my commitment with each core value as listed

2. Expressing each value commitment out loud to myself, and ideally to another I feel responsible to

3. Continuing to reassess my commitment of my pledge to myself and the others I serve, through regularly scheduled written and spoken self-reflection

4. Committing to maintain regularly scheduled, mutually agreed upon reviews with those I serve

5. Signing my name as my promise to live by each value

This is my word.

Signed

(It is recommended that a yearly day of celebration be initiated to honor every citizen's living by their sacred pledge).

~

My hope...my dream....my wish......my prayer.......is that this aforementioned Vision of the Sacred....Love.... All One ....leads us to a .....Re-Set .....a Re-Do... So that we use all we've learned in the past four hundred years.....to re-birth....re-create a nation that truly is for All The People. (And maybe, since it was built by enslaved Africans for the wealthy white hetero men in power, we can even consider renaming The "White" House, The All One House. or something like that).

<u>If all the policies and laws of our democracy, met these aforementioned values and standards, then "We The People" will truly mean "All The People"......and "All Life.</u>" May it Be So.

~

THE MASTER'S
TOOLS
WILL NEVER DISMANTLE
THE MASTER'S
HOUSE.

AUDRE
LORDE

~

Letter to the President

The following was submitted to the President, Vice President, and selected members of Congress, with hope that Loving and Courageous action will be taken:

February 7, 2022

Dear President Biden, Vice President Harris, and Select Members of Congress,

We are writing personally to you, because you have demonstrated a willingness to speak out against racial injustice, and to urge you to consider initiating and being part of an Inaugural Truth and Conciliation Commission. It is our prayer and hope that you will review the attached proposal toward the healing of our country...and the world.

Please know what we are sharing is not intended to guilt trip you or us or anyone ...as we weren't taught about this information. However...<u>now</u> is the time to fully educate ourselves and each other if we want to mend the problems and divides that have plagued us from the beginning of our country.

Given your wisdom, compassion and courage, we believe you are perfectly positioned at this moment in time, to do something no other President or members of Congress have been willing to……that is to finally speak the full truth about the white supremacy value system that founded this country and still runs it. And that this system was started by and still run by ….predominantly wealthy, white heterosexual men, who have never taken full responsibility and accountability for the incredible hurt and damage inflicted on….<u>All</u> of us. We strongly believe that this needs **to be powerfully and publicly stated as part of an ongoing, country-wide Truth and Conciliation / Transformation process**, if we have any hope of real sustainable healing and change.

Since we have never adequately addressed the **root cause** of all the "isms" sexism, racism, homophobia, classism, xenophobia and on and on, it is no wonder that so many wonderful efforts have fallen short. We keep circling round and round, mentioning subtly or rapidly glossing over, rather than digging into the roots. If the poison has not been fully acknowledged and addressed, and vetted by those who have been oppressed…..in an on-going way …how do we expect policies and programs to effect sustainable change? With love, and fairness, and justice for **All** life (including the animals and the Earth), as the carefully monitored, guiding principles, there is real hope for sustainable trust and healing to be forged.

This would be a wonderful, impactful example set by you, and especially other white male leaders, in government and all other systems, which could be followed throughout our country...and the world.

So we are asking, you to please consider the attached article on "Mental Health and Wealthy White Hetero Male Supremacy...the Dilemma", and proposal, entitled "All Life is Sacred", and some related poems, as we think the theory and blueprints for healing are contained within them.

Thank you for your service and for who you are.
Sincerely,

Garett Reppenhagen	Roberto Schiraldi
Executive Director	Member
Veterans For Peace	Veterans For Peace

Attachments:
Introductory Poems - "white supremacy culture", "confederate flags in the Capital Halls", "white supremacy healing", "Climate Crisis and militarism". "One Brave Man"
Background Article - "Mental Health and Wealthy White Hetero Male Supremacy...the Dilemma"

Proposal - "All Life Is Sacred"
Recommended Members of Inaugural Truth and Conciliation/Transformation Commission

P.S. President Biden actually replied with a hand written note of thanks, saying he liked the poetry, and included one of his favorite poems of hope.

~

Outline for Core Curriculum Course

(The following is part of a proposal submitted to the New Jersey State Licensure Board, and the American Counseling Association in hope of ensuring that all professional counselors receive training in what we believe to be essential ingredients for addressing racial trauma. It was also used as the basis for a number of trainings, including one for the leadership of the North Atlantic Region of the American Counseling Association. (We believe it is unethical to not require that every counselor receive the information and personal integration that this type of training offers (please see section IIB 16 below for more on the ethical consideration).

This material is also relevant, **highly recommended and easily adaptable** for all institutions including; **High school and Higher Education** (especially Graduate Counseling Programs,) Government, Military, Police, Corporations, Prisons, Banks, Churches, Families, and Leadership Training).

> **Accurate History. Acknowledging key components of US history ("Racial Literacy"), provides a context and implications for addressing current individual and societal problems.** It is incredibly challenging to heal intergenerational racial trauma wounds of the past and present without first acknowledging some of the painful historical truths that were part of the founding of our country and are likely continuing to contribute to our current racial problems. Having a

"common understanding" of our accurate history is essential for sustainable healing.

IA. Genocide of Native Americans

Mass killing of Native Americans

Doctrine of Discovery, God's Anointed Ones, Manifest Destiny, as the ideology to justify Europeans' economic advancement **at all costs**

Seneca form of tribal government used as model for democracy

Removal Act of 1830

Intentional exposure to smallpox

Forcing children to separate from their families and communities by sending them to boarding schools and forced to assimilate

Slaughter of buffalo, starvation, and social and cultural disintegration of many Plains tribes.

Economic disintegration of many Plains tribes due to many treaties violated by US government.

Present situation: Still mostly confined to reservations, poverty, poor education, and minimal job opportunities.

Contributions by Native Americans to American culture

Understanding and acknowledging resulting intergenerational trauma is necessary for healing and trust building to begin. (Truth and Reconciliation Commission of Canada, South African Truth and Reconciliation Commission as examples)

There are many Native Americans in NJ, for example the Nanticoke Lenni-Lenape. Many African Americans have Native American ancestry, many multiracial people have Native American ancestry. Territorial acknowledgment, created in consultation with Indigenous people, is one way to demonstrate respect for their history and culture when hosting training programs and organizational meetings.

Bibliography: 1. NJ Dept. Of Education. Amistad Commission http://njamistadcurriculum.org 2. Zinn Education Project http://zinnedproject.org 3. *An Indigenous People's History of the United States (Revisioning American History)* by Roxanne Dunbar Ortiz (2014) 4. Truth and Reconciliation Commission of Canada http://www.trc.ca/websites/trcinstitution/index.php?p=3 5. *"Race: The Power of an Illusion"*, Three-part educational documentary on PBS, by California Newsreel, www.youtube.com/watch?v=Y8MS6zubIaQ. 6. *Decolonization Is Not A Metaphor,* Tuck, E., & Yang, K.W., Decolonization: Indigeneity, Education & Society, Vol. 1, No. 1, 2012, pp. 1-40.

IB. Enslaved Africans Viewed as Property, Not as People, and the Legacy of Slavery

Africans intentionally chosen to be kidnapped because they were strong, smart, and master farmers

Major difference between Africans, the only group forced to come here, and all other immigrant groups that came to America by choice.

Original U.S. Constitution was written for land owning, white males

Atrocities of slavery: forced labor, torture, murder, rape, body mutilation, breakdown of family and culture.

Identifying the U.S. Presidents who owned slaves.

- Civil War / real reasons for Civil War, economic/freeing the slaves (slave labor undergirded the U.S economy.). As slaves were freed, the Ku Klux Klan and other hate groups grew in number to "protect their rights". Thus, the original manifest destiny doctrine, combined with fear of the "angry Black man" can be viewed as key components to developing implicit or unconscious bias towards African Americans as well as many other groups (such as Asians, Latinos, other immigrant groups, LGBTs, women, people with disabilities), which may be perceived as a threat to white supremacy. An effective colonization strategy used to control the masses is by pitting them against each other.
- Policing system started, with bounty hunters to retrieve run away, enslaved Africans. This has clearly contributed to implicit biases within the system, seeing Black men as threatening/ fear of "angry" Black men "raping our women" etc., often leading to extreme violence against Black men, and grossly inflated incarceration of generations of Black mothers' sons and Black children's' fathers.
- Post-emancipation and Reconstruction sharecropping virtually re-establishing the plantation system and Jim Crow laws (loitering, etc.) resulting in chain gangs (penal slavery).
- Voting rights /still an issue for people in their communities and for those who get out of jail.
- Mass incarceration and killing of young Black men.
- Contributions by African Americans to American culture, regularly misappropriated by other cultures in power. Acknowledging and celebrating resiliency

and joy of the resistance experience of family and social traditions of African Americans and other marginalized groups, and their huge impact on the arts, music, dance, poetry, fashion, cuisine, language arts, religion/spirituality, athletics, education, science, medicine, science, industry, business, government, etc.

Understanding and acknowledging resulting intergenerational trauma derived from this status of oppression, which is necessary for healing and trust building to begin. (Truth and Reconciliation Commission of Canada, South African Truth and Reconciliation Commission as examples).

*Understanding how the above two sections contributed to oppressive treatment of and minimizing contributions by Asian, Latinx and other people of color, and other groups to American culture i.e., manifest destiny and the U.S. Constitution for land owning white males, everyone else excluded.

Bibliography: 1. NJ Dept. Of Education. Amistad Commission http://njamistadcurriculum.org 2. Zinn Education Project http://zinnedproject.org. 3. *Slavery by Another Name: The Re-Enslavement of Black Americans from the Civil War to World War II* by Douglas Blackmon. (2009) 4. *The New Jim Crow: Mass Incarceration in the Age of Colorblindness* by Michelle Alexander (2010) 5. Hoover, Jania, *"Don't Teach Black History Without Joy"*, Education Week, February 2, 2021,

*Skills. Obtaining the knowledge and skills to comfortably discuss and address White Supremacy

Culture, White Privilege, Intersectionality Theory, Implicit Bias, Racial Microaggressions, Broaching, Non-violent communication, and Healing From Racial Trauma, are essential ingredients for addressing racial problems.

IIA. White Supremacy Culture

1. Understanding white supremacy culture and white supremacy cultural values and how they contributed to above sections A and B). **System of power, wealth and control that founded this country and still runs it.... a hierarchy of human value, orchestrated by predominantly wealthy / land owning white heterosexual males, where inferiority and superiority are based primarily on race** .... a system which cleverly manipulates and pits all races, ethnicities, women, poor and middle-class white men, and all other groups, against the other, each competing for an equitable piece of the pie.

2. **Characteristics* and values of white supremacy culture – (Stemming from principles of "God's Anointed Ones, Doctrine of Discovery, Manifest Destiny)...leading to elitism, entitlement, superiority, hyper masculinity, competition to be #1 at all costs, no matter who you need to walk over to get there, success, power, wealth, especially material wealth, "perfectionism, sense of urgency, metered goals (often leading to "It's never quite good enough"/ "I'm never quite good enough", multitasking,** worship of the written word (if it's not in a memo, it doesn't

exist), there's only one right way, "quantity over quality, defensiveness, paternalism, those in power make decisions for the group, either/or thinking, good/bad, right/wrong, us against them, power hoarding, little if any value around sharing power, fear of open conflict, individualism rather than working as team, progress is bigger, more, objectivity, belief that there is such a thing as being objective and "neutral", scapegoating those who cause discomfort*

3. ***** <u>**Unhealthy/"Toxic" Masculinity** (which hurts us all) / Intersection of sexism and racism</u>, patriarchy, capitalism, and the importance of a feminist/racial equity perspective for counseling. (Books by Jackson Katz, i.e., "Man Enough", "Leading Men", "The Macho Paradox", National Organization for Men Against Sexism). *This may be the most important factor, which significantly contributes to, and underlies all the other factors of racism.* <u>*The unhealthy wealthy white hetero male supremacy cultural values*</u> *which founded this country and still run it...have never been adequately acknowledged or examined...i.e., wealthy white heteromaledomination/patriarchy, power, wealth, control, success, elitism, entitlement, competition to be #1 at all costs, no matter who we need to walk over, and on an on. Until the wealthy, white hetero male supremacy power elite take responsibility for the values which run things, and commit to making it right....we will keep running around in circles, putting temporary band-aids on the problems. This includes the mental health and social service systems, which diagnose "problems", in order to keep the power structures in*

place, rather than insist on effectively treating the root causes of the problems in the first place.

4. ** **Dwelling in a culture of fear.** The aforementioned unhealthy masculinity leads to dwelling in a culture of fear, where there are so many threats to "safety" and "security"….i.e., terrorism, war, climate change and environmental disasters, stress related dis-ease / pandemics, civil unrest/violence, economic volatility and on and on. 'The Fear Of Women', by Austrian Psychiatrist Wolfgang Lederer, examines how men, down through the ages have both loved and feared women (given women's incredible power to give birth, and their connection to the mother earth..ie. burning witches etc). **Fear** of being **vulnerable,** losing "control", of not being good enough, can lead to being macho, as a cover-up., see presidential debates like two school boys posturing in the school yard….what a terrible, terrible example for our young boys. It begs the question, is fear of being gay (I.e, two men actually being vulnerable, tender and caring with each other), a core root of so many of our problems? *(Please see the amazing video 'The Wisdom of Trauma', by Dr. Gabor Mate, which explores the greatest trauma of being alienated from ourselves, and the important work of Brene Brown on being vulnerable, including her Ted Talk, 'Power of Vulnerability', and her audio CD, 'Teachings on Authenticity, Connection and Courage')*. This fear of being vulnerable, and the loving courage it takes to be so, may well hold the key to healing for us all, (Schiraldi, R., 'A White Man on the REZ, "Higher" Education in a Culture of Fear", Princeton Comment.com, April 29, 2013

5. ****Antidotes* to white supremacy cultural values / Creating a culture of Healthy Values**- Choosing to teach and model and live by agreed upon values which **serve All** Life - Starting with our relationship with ourselves and all others, in all of our institutions... families, schools, (K-Post Grad as core mission), churches, health care/MH, government, business and industry, law enforcement, military, prisons, and on and one, asking for commitment to core values in all life. **Teaching the core value of All Life is Sacred. Sacred, meaning...highly valued, worthwhile, important, precious, cherished, worthy of love, dignity, respect, awe and wonder. I am no better no worse than that insect, that plant, that tree, that animal, that child, that woman, that man. All Life is Sacred. Teaching about Love* as our most important value - including such essential ingredients as impeccable integrity, courage, humility, vulnerability, commitment, responsibility, cooperation, sharing, generosity, kindness, consideration, nurturance, support, encouragement, dependability, gentleness, strength, service, respect, accountability, equity, fairness. (see bell hooks,'All About Love'***)**. Develop a culture of appreciation, i.e., especially in work places** - by establishing "realistic work plans/time frames, set goals of diversity and inclusion, understanding the link between defensiveness and fear of losing power, losing face, losing privilege etc., including process or quality goals in planning, to get off the agenda to hear others', concerns, learn how others communicate, come up with other ways to share information, accept there are many ways to do something and be willing to learn about other's

different cultural ways of doing things, include people who are affected by decisions in the decision making process, push to come up with more than two alternatives take breaks, breathe, avoid making decisions under extreme pressure, include sharing power in your mission statement, make sure there is understanding that a good leader develops the power and skills of others, and that change is inevitable, revisit how conflict is handled, and see how it might be handled differently, create culture of group solving as value statement, create 7th generation thinking, how actions will affect seven generations from now, be open to understanding different world views, deepen understanding of racism and oppression and see how personal experience fits into the larger picture", (** *"white supremacy culture", Changework, Dismantling Racism Workshop*, Tema Okun).

Bibliography. Ruiz, D. M., 'The Four Agreements', Howe, M. and Howe, L. 'Values Clarification, *hooks, b., 'All About Love', Human Development Program, Charles, M, *"We the People"* (U-Tube Video), Wilson-Shaef, A. 'When Society Becomes The Addict', Hillman, J., Ventura, M., 'We've Had A Hundred Years of Psychotherapy and the Worlds Getting Worse', Schiraldi, G., 'Self-Esteem Workbook', Schiraldi, G., 'Complete Guide for Resiliency', Schiraldi, R. *"white man on the REZ"* (Journal Article), Schiraldi, R., 'Healing Love Poems for white supremacy culture: *Living our Values'*, Schiraldi, R. 'Unexpurgated* Racial Justice Poetry with Healing Meditations'.

6. <u>White supremacy cultural values create racism, sexism, homophobia, xenophobia, classism</u> and all the

other issues which divide and **instill mistrust** among the groups, all for the control of the few wealthy white men in power. For example, this **fear** and **mistrust** is an especially effective way of influencing working-class decision-making regarding union activities and voting. Bibliography. Lorde, A., 'The Master's Tools Will Never Dismantle The Master's House', Penguin Random House, 2017.

7. The ongoing **mistrust** can create even more animosity during challenging times, i.e., after 911 Arab and Muslim people were often targeted, and many wore American flags to avoid attacks and being accused of being foreigners or unpatriotic. **"It is not our differences that divide us. It is our inability to recognize, accept, and celebrate those differences."– Audre Lorde. The popular notions of "melting pot", and "assimilation" which seem so attractive on one hand, can also be viewed as effective tools to lessen unique ties to cultural traditions, thus making it easier to control and manipulate.**

8. U.S. Constitution excluding Women, Native men and counting Black men as $3/5^{th}$ human (to control voting power). See 'We The People', Ted Talks by Mark Charles, Navaho.

9. "We cannot change what we don't acknowledge." - James Baldwin. "When we lack a common understanding of our history, it is much more difficult to come together." - Mark Charles

10. It is curious how a country founded upon revolutionary spirit, quickly adopts a strong "nationalistic fervor", and accuses protesters of being un-American, often leading to violent confrontations ire., Black Lives Matter and Charlottesville. **"Revolution is not a one-time event. It is becoming always vigilant for the smallest opportunity to make a genuine change in established, outgrown responses; for instance, it is learning to address each other's difference with respect.** "Audre Lorde

11. Considering the implications of racial breakdown of people who control our institutions, the vast majority being white hetero males. **White supremacy cultural values lead to systemic racism,** which permeates all our institutions- government, education, religion, corporations, banks, police, military, sports, health care, and mental health. **"The master's tools will never dismantle the master's house. They may allow us to temporarily beat him at his own game, but they will never enable us to bring about genuine change."** – Audre Lorde

12. **Four hundred plus years in a white supremacy culture has yielded some improvements.** Yet racism and inequity still flourish because we haven't gotten to the root cause. The root cause is that "All Life, is not and never has been, seen as sacred, valuable, worthwhile. If it were, we wouldn't have the need for a curriculum about white supremacy culture. Yet hope burns eternal. And the "struggle" goes on. "If there is no struggle, there is no progress. Those who profess

Men and Racism: The Healing Path | 197

to favor freedom, and yet depreciate agitation, are men who want crops without plowing up the ground. They want rain without thunder and lightning. They want the ocean without the awful roar of its many waters. This struggle may be a moral one; or it may be a physical one; or it may be both moral and physical; but it must be a struggle. <u>Power concedes nothing without a demand.</u> <u>It never did and it never will</u>."
— Frederick Douglass, <u>Frederick Douglass: Selected Speeches and Writings</u>

13. <u>A Truth and Conciliation Process</u> (conciliation rather than re-conciliation, which implies that there once was harmony), like those in Africa, Australia, Canada, and some related efforts in this country, have proven to be helpful in at least beginning the process of healing deeply embedded wounds. The process entails acknowledgment of injustices and atrocities, sincere, heartfelt amends, and commitment to sustainable change, vetted by those who have been oppressed and traumatized, often inter-generationally. Of course, trust takes time to build, and then, only, if the ones in power keep to their word.

14. <u>Impact on mental health issues</u> with people of color and white people, from living in white supremacy culture. Mental health can be viewed as the result of a balance of mental, emotional, physical, spiritual, social aspects of our lives. Racism and it's skewed values are traumatic, the antithesis of good mental health, and extremely hurtful to our mental, emotional, physical, spiritual, and social health. The intersection of corona virus and nationwide protests against racism can be

viewed as the simultaneous result of a systemic history of dis-ease disorders...because we are all dwelling in a disordered culture of fear, where our very humanity and sense of self-worth depends on so many things outside of our control. This leads to incredible stress related illness and mistrust of ourselves and each other, especially those who we perceive as "different", and a threat.

Bibliography. 1. Bryant-Davis, T. (2007). *Healing requires recognition: The case for race-based traumatic stress.* The Counseling Psychologist,35,135-142., 2. Hemmings, C. Evans, A. *Identifying and Treating Race Based Trauma in Counseling*(2018), Journal of Multicultural Counseling and Development, Vol 46. Issue 1 20-39IV, 3. Telusma, B., *Do We All Have PTSD? Mental Health in an age of racial terror.* Grio, March 27, 2018.

15. Implications of white supremacy culture on counseling profession i.e., many counselors over diagnose clients of color, and miss racial trauma diagnosis. It is essential that counselors understand how white supremacy culture leads to racial trauma and resulting stress related dis-eases which affect us all. And the concept of "disorder", might well be replaced with adjustment reaction to a dis-ordered society (a la Anne Wilson Schaef's seminal book *"When Society Becomes an Addict."*

16. ****Ethical implications of lack of necessary anti-racism training for all counselors**. A study published in the Journal of Multicultural Counseling and Development investigated "**Are counselors really addressing the issue of race-based trauma?**" It found that the ***majority***

of counselors in the United States are <u>not prepared to identify and treat race-based trauma</u>, which often results from racial harassment, discrimination, violence, or experiencing institutional Racism. Chronic racism and discrimination can lead to a wide variety of psychological problems, including denigration of one's sociocultural in-groups, feelings of helplessness, numbing, paranoid like guardedness, medical illness, anxiety, fear, and the development of posttraumatic stress disorder. (Hemmings, C., Evans, A., Identifying and Treating Race-Based Trauma in Counseling, (2018), Journal of Multicultural Counseling and Development, Vol. 46, Issue 1, 20-39IV). The importance of ensuring the highest level of training in anti-racism work, healing trauma work, and how white supremacy impacts the counseling profession, cannot be emphasized enough. **To have a workforce of counselors who are inadequately trained on these issues, can end up re-traumatizing clients, and will certainly increase mistrust in the counseling profession.** It is unethical and irresponsible to license counselors who haven't received training on the core issues addressed here. This is one of the most vital issues of our time. (Please see these important statements and articles: 1. "APA Apology To People of Color for APA's Role in Perpetuating, Promoting, and Failing to Challenge Racism, Racial Discrimination, and Human Hierarchy in U.S." Resolution Adopted by the APA Council of Representatives on October 29 2021, 2. "Culture Centered Counseling", by Lindsay Phillips, Senior Editor of Counseling Today, ACA, November 22, 2021, 3. ACA Statement on Anti-Racism, by ACA Advisory Council, June 22, 2020).

17. <u>Importance of counselors doing their own inner work</u> on the effects of dwelling in a white supremacy culture, and its impact on our humanity. This is **lifelong learning** which requires a willingness to be vulnerable and continuing to discover, unpeel, acknowledge our implicit/unconscious biases, white fragility/defensiveness/guilt. We were never taught this information, so how would we know. This is difficult, messy work. And none of us will do it perfectly. We will make mistakes. It's about acknowledging our mistakes, considering why it happened/where we learned it, apologizing when appropriate, and coming up with a reasonable plan to prevent a repeat, and then committing to carry through on our commitment. This is what leads to real growth making us feel more confident if we persevere, with gentleness and compassion for ourselves and each other. It makes us much more likely to create trust with our clients from marginalized backgrounds.
(See these important books for doing inner work, 1. "<u>How to Be An Antiracist</u>", by Ibram X. Kendi, Random House (2019). 2. <u>Diversity in Clinical Practice</u> by Lambers Fisher, PESI Publishing (2019) 3, <u>Promoting Cultural Sensitivity in Supervision</u>, by Kenneth V. Hardy, and Toby Bobes, Routledge (2017) 4. <u>Mindful Of Race</u> by Ruth King, Sounds True (2018) 5. <u>The Racial Healing Handbook</u>, by Anneliese A. Singh, New Harbinger Publishing (2019) 6. "<u>Me and White Supremacy</u>", by Layla F. Saad, Sourcebooks (2020) and 7. "<u>White Fragility</u>", by Robin DiAngelo, Beacon Press (2018)).

18. <u>Role of white supremacy cultural values leading to white privilege and racism</u>. Understanding the

intersection of white supremacy, racism, and white privilege. For example, when one group is "supreme" (from God's anointed ones, doctrine of discovery manifest destiny), all other groups are not. This sets us all up for discrimination and racism. And, a supreme" group will automatically have benefits and privileges that other groups will not have...thus white privilege.

Bibliography: 1.*"What is WhiteSupremacy"* by Elizabeth "Betita" Martinezwww.CollectiveLiberation.org 2018.
2. *"No, I Won't Stop Saying White Supremacy"* by Dr. Robin DiAngelo Good Mens Project; August 12, 2018. 3. *"The History of Patriarchy"* by Ellie Bean Medium.com, November 11, 2018.
4. *'Post Traumatic Slave Syndrome'* by Dr. Joy Degruy, Uptone Press 2005. 5. *My Grandmother's Hands: Racialized Trauma and the Pathway to Mending Our Hearts and Bodies* by Resma Menakem Central Recovery Press, Las Vegas, NV, 2017. 6. *"White Supremacy Culture"*, by Tema Okun, Changework Racism Workbook, Oakland, Calif., 1999. 7. *"White Supremacy Culture"*, New Resolution, National Education Association Report, July 2018.

IIB. White Privilege

1. Definition of white racial privilege (and how it contributed to above sections I.A and B). Advantages and benefits people have because they are white (see list of benefits in McIntosh article cited below).
2. Definition of racism. Racial prejudices exercised against a racial group by individuals and institutions in a position of power.

3. Role of institutional power. (Internment camps for Japanese Americans as example of treatment of other marginalized groups).
4. Effect of "unhealthy" white heterosexual male values (superiority, power, wealth, control, elitism, entitlement, success, competition to be number one at all costs, defining beauty by European standards, etc.).
5. Internalized oppression of people of color. Accepting and internalizing stereotypes and myths exposed to.
6. Colorblindness. While seen by some as well-intentioned, denies the reality of existing racism.
7. What to do?
   a. It is important to acknowledge our privilege and take responsibility to speak out, support, and advocate, if there is to be substantial, sustainable change. This is not about guilt or shame; it is about acting. Complicit bias asserts "if we're not part of the solution, we're part of the problem".
   b. **<u>Teach core value that All life is Sacred -valuable, worthwhile, important, worthy of love, dignity, and respect, that all individuals have worth, regardless of grades, money, possessions.</u>** This is a fundamental Native American teaching.
   c. <u>Teach importance of Dr. MLK's speech, "I have a dream that my four children will one day live in a nation where they will be judged not by the color of their skin but by the content of their character."</u>
   d. **Demonstrate an on-going commitment to learning about and becoming comfortable with discussing White Supremacy Culture, White**

Privilege, Intersectionality Theory, Implicit Bias, Microaggressions and Broaching when appropriate.

e. Appreciating the trap created by depending on a hierarchical notion of life. This limiting paradigm labels cultures as "superior" or "inferior". In reality, human experience within culture cannot be placed in this hierarchy. Note the powerful concept of life as a circle rather than a pyramid. When counselors use the circular perspective, our work with clients is deeply transformed.

Bibliography: 1. *Teaching Tolerance/Teaching the New Jim Crow* by the Southern Poverty Law Center. http://www.tolerance.org/publication/teaching-new-jim-crow 2. *"Unpacking the Invisible Backpack"* by Peggy McIntosh, Wellesley College 1989. 3. *Teaching for Change.* http://www.teachingforchange.org 4. White Privilege Conference Course Curricula. NJ Dept. Of Education. Amistad Commission http://njamistadcurriculum.org 5. *The Great White Elephant: A Workbook on Racial Privilege for White Anti-Racists* by Pamela Chambers and Robin Parker. Beyond Diversity Resource Center, 2007 http://beyonddiversity.org/books, 6. *"Under-standing Internalized Oppression"*, by Teeomm K. Williams, Univ. of Mass. 2012. 7 *"Being Colorblind does not offset innate advantages of White Privilege"*, by Robert Jensen, Kansas City Business Journal, January 5, 2001, p.20. 8. *The Self-Esteem Workbook* by Glenn R. Schiraldi, pp. 29-37, New Harbinger, 2001. 9.*"A Journey through Alienation and Privilege to Healing,"* by Roberto Schiraldi, White

Privilege Journal 2013. *http://www.wpcjournal.com/article/view/6457.*

IIC. Intersectionality Theory

1. Examination of factors of identity (gender, race, class, economic status, ability, sexual orientation, etc.), which intersect on multiple and often simultaneous levels
2. Recognition that people can be privileged in some ways and not privileged in others, and there may be limited awareness of one identity and not others
3. Multiple identities and systems of oppression at work and their impact on our counseling diagnostic tools
4. Importance of increasing awareness and not reinforcing deficit models of cultural groups.
5. Lack of experience in interacting with individuals from different groups can lead to over- or under-emphasis on cultural factors (**"Broaching"**).
6. Important never to assume, but to ask, if appropriate, about individual identities, (**"Broaching"**).

Bibliography: 1. *Race, Class, Gender: An Anthology* by Margaret L. Andersen and Patricia Hill Collins (8th ed. 2012) 2. *"Demarginalizing the Intersection of Race and Sex,"* The University of Chicago Legal Forum 140:139-167 (1989), by Kimberly Crenshaw 3. *Black Feminist Thought: Knowledge, Consciousness, and the Politics of Empowerment,* by Patricia Hill Collins. (2008), Messner, M. 'Unconventional Combat: Intersectional Action in the Veteran's Peace Movement', Oxford Press, 2021.

IID. Implicit Bias

1. Definition. Prejudices that are unknown to the conscious mind, involuntarily formed, and usually

denied. Developed over a lifetime through exposure to direct and indirect messages, early life experiences, the media and news programming. We all have them.

2. Examples of implicit bias (based on extensive research cited by Kirwan Institute below and Harvard University below, view and take IAT, Implicit Association Test).).

2a. In emergency rooms, whites are pervasively given stronger painkillers than Blacks or Hispanics.

2b. Everyone is susceptible to implicit bias, even people who believe themselves to impartial or objective, such as judges.

2c. College student video game participants are more likely to "shoot" when the target is black.

Bibliography. 1. *The State of the Science: Implicit Bias Review 2013*, Kirwan Institute for the Study of Race and Ethnicity, Ohio State University, 2. IAT, Implicit Association Test, https://implicit.harvard.edu/implicit. 3. "Across *America, Whites Are Biased and They Don't Even Know It"*, by Chris Mooney, Washington Post, (2014). 4.*"Blindspot: Hidden Biases of Good People"*, by Banaji, M., and Greenwald, A., Random House, 2013.

IIE. Racial Microaggressions

1. Definition. Everyday insults, indignities and demeaning messages usually unintentionally sent to people of color by white people who are unaware of the hidden meanings embedded in the communication.

2.Examples of microaggressions.

a. Myth of meritocracy which asserts that race, gender, class, sexual orientation does not play a role in life success.

b. "Color blindness" (Denying a person of color's racial or ethnic experience; the implicit message is "They should just assimilate").

c. Assumptions (Asian Americans and Latino Americans are assumed to be "perpetual foreigners;" or "inheritantly inferior", based on biased assumptions of intelligence based on race, gender, or perceived abilities).

<u>Bibliography</u>. "Racial Microaggressions In Everyday Life: Implications for Clinical Practice," by Derald Wing Sue, *American Psychologist*, pp. 271-286 (2007) 5. *Diversity and Oppression,* Graduate MSW Course taught by DuWayne Battle, PhD, Rutgers University/MSW Culture Competence Certification Program.

IIF. Broaching

1. Definitions. A consistent and ongoing attitude of openness with a genuine commitment to explore issues of diversity. Addressing and/or responding to racially offensive comments, attitudes, and behaviors.
2. Examples of broaching styles:
    a. Avoids broaching. Regarded as unnecessary. Defensive when asked to broach.
    b. Broaches sporadically. Vacillates due to discomfort, lack of skill, and concern about negative reactions from others.
    c. Consistently broaches. Integrated and congruent with commitment to social justice.

<u>Bibliography</u>. 1.. *Broaching the subjects of race, ethnicity, and culture during the counseling process,* by N.L. Day-Vines et al, Journal of Counseling & Development, 85, pp. 401-409 (2007) 2. *Helping school counselors*

successfully broach the *subjects of race, ethnicity, and culture during the counseling process,* by N.L. Day-Vines & T. Grothaus, ASCA Conference, Chicago (2006).
3. *Promoting Racial Literacy in Schools*, by Howard Stevenson (2014).

IIG. Non-violent Communication / Compassion Focused Therapy

1. Definitions. Non-violent Communication and Compassion Focused Therapy are two powerful approaches to assist us in learning how to give ourselves and others empathy and compassion as we address these often very painful issues.
2. Examples.
2a. Work internally to acknowledge and feel compassion for own feelings and needs. Making a request of self. "Would I be willing to give myself respect, support?"
2b. Express how I am to another without blaming or criticizing. "Would you be willing to help me understand what you're experiencing?"
Bibliography. 1. *Non-violent Communication*, by Marshall Rosenberg, Puddle Dancer, 2015, *Healing Across Differences*, by Dian Millian, Puddle Dancer, 2012, *Compassion Focused Therapy*, by Paul Gilbert, Routledge, 2010.

IIH. Healing From Racial Trauma

1. Definition - Race-based traumatic stress is an emotional injury that is motivated by hate or fear of a person or group of people as a result of their race, a racially motivated stressor that overwhelms a person's

capacity to cope, a racially motivated, interpersonal severe stressor that causes bodily harm or threatens ones' life integrity, a severe interpersonal or institutional stressor motivated by racism that causes fear, helplessness or horror. Intersecting identities can result in multiple traumas and forms of oppression. Racial trauma can be viewed as intergenerational, that is, passed on from generation to generation, thus compounding the trauma.

2. Common reactions to Racial Trauma- shock, hopelessness, anger, numbness, rage, disbelief, grief, panic, preoccupation, guilt, ADHD, loneliness, overwhelming despair, flashbacks, uncontrollable tearfulness, fatigue, mistrust, high anxiety, low self-worth, increased use of alcohol and other drugs, headaches, pains, overeating, gastrointestinal disorders, hypertension, compromised immune system, pervasive sense of loss, and loss of feeling safe and others.

3. Approaches to Healing- <u>Racialized Trauma Therapy should only be offered by well trained professionals who specialize in racial trauma and who have extensive experience and supervision in providing this support. Counselors who have not done been well trained and not done their own inner anti-racism work can re-traumatize clients.</u> Feeling trust between client and therapist is essential and takes times to build. Racial trauma therapy can include acknowledging events, feelings, memories, being believed/ feeling heard and understood, having reactions and symptoms normalized, unburden family secrets, deep grief and rage work, rehearsing future safety measures, meditation, mindfulness meditation, keeping journal of self care / self-love practices, affirmations, acknowledging resilience and strength,

spiritual practices, clear boundaries, accurate education, identifying needs like comfort, reassurance, compassion, respect.

4. <u>Rage work</u> - members of oppressed groups (including professional counselors and graduate counseling students), often need space to verbalize their rage at the countless injustices they have dealt with almost all their lives. For someone with no similar history, this is extremely challenging to witness and facilitate without rushing to tone down. This kind of emotion is also present in women & men who have been sexually abused.

<u>Bibliography</u>. key authors on the psychology of oppressed people:

> Dr. Frantz Fanon. He crafted the moral core of decolonization theory as a commitment to the individual human dignity of each member of populations typically dismissed as "the masses". His book is a must read "Black skins, white masks" (1952). 2) Paulo Freire who taught us so much of what it takes to help clients who are persistently oppressed (his theory is "The pedagogy of the oppressed").\Freire, P. (1970). Pedagogy of the oppressed. New York, NY: Continuum. 3) Dr. Lillian Comas-Diaz, former APA president whose articles on the psychology of Latin women is fundamental to understanding how to work with women who come from oppressive circumstances in a way that affirm them rather than just deleting their symptoms. Comas-Diaz, L. (2006). LATINO HEALING: The integration of ethnic psychology into psychotherapy. Journal of Psychotherapy: Theory, Research, Practice, Vol. 43 (4):436–453. (2)Comas-Diaz, L. (2011).

Multicultural approaches to psychotherapy. In J. C. Norcross, G. R. VandenBos, & D. K. Freedheim (Eds.), History of psychotherapy: Continuity and change (2nd ed., pp. 243–267). Washington, DC: American Psychological Association. 4) Scholar Dr. W.E.B. Dubois, Dubois, W.E.B. "The Future of the Negro Race in America." DuBois on Reform: Periodical-Based Leadership for African Americans. Ed. Brian Johnson. Lanham: AltaMira Press, 2005. 160-171. He coined the construct of **double consciousness.**

Other Important Work on Healing Racial Trauma:
1. Bryant-Davis, T. (2007). *Healing requires recognition: The case for race-based traumatic stress.* The Counseling Psychologist,35,135-142., 2. Hemmings, C. Evans, A. *Identifying and Treating Race Based Trauma in Counseling*(2018), Journal of Multicultural Counseling and Development, Vol 46. Issue 1 20-39IV, 3. Telusma, B., *Do We All Have PTSD? Mental Health in an age of racial terror.* Grio, March 27, 2018, 4. DeGruy, J., *Post Traumatic Slave Syndrome,* Uptone Press (2005), 5. Menakem, R., *My Grandmother's Hands,* Central Recovery Press (2017). 6. *Scientific American, The Science of Overcoming Racism: What Research Shows and Experts Say About Creating a More Just and Equitable World.* Special Collectors Edition, Summer 2021. Volume 30, Number 3.

Summary. Acknowledging accurate history, and its connection to white supremacy culture, white privilege, racism, and intersectionality theory can assist us in understanding past and current racial struggles and pave the way to practice broaching and non-violent communication with our unintentional microaggressions, and the unconscious biases which may trigger them. With these understandings and internal work, counselors will be much more able to provide compassionate and supportive counseling for all their clients, especially those experiencing racial trauma.

© 2020 Schiraldi (Williams, Rodriguez)

I Have A Dream*

-That we men, especially us white, heterosexual men,** will commit to uncovering all vestiges of white supremacy inside ourselves and this culture, not out of guilt or fear, but out of love and compassion, and justice

-That we men will start choosing Love over Fear in all our decision making

-That we men will <u>demand of ourselves</u> to stop hurting ourselves and others

-That we men will work on our own insecurity, so we will stop putting down others because they're Native, or Black, or Muslim, or Asian, or Gay, or just different

-That we men will stop being bullies with verbal, and physical threats or intimidation

-That we men will stop Choosing To Be out of control with alcohol and other drugs

-That we men will stop beating up other men to prove our masculinity

-That we men will stop raping drunk or drugged women, or any women

-That we men <u>will</u> support each other in standing up and speaking out against racism, sexual assault, sexism and homophobia

-That we men will commit to learning how to express our anger in healthy, non-violent ways

-That we men will seek help to heal the pain of our childhoods

-That we men will choose to carry ourselves with respect and dignity all the time

-That we men will be courageous enough and strong enough to be gentle and kind with ourselves and all others

-That we men will hold ourselves accountable for all of our thoughts, words and behaviors, all of the time

-That we men will remember that the real truth, happiness and serenity lies within
-That we men will choose, <u>right now</u>- to learn to love and nurture <u>ourselves,</u> physically, mentally, emotionally, spiritually, and socially, and make <u>this</u> our way of life

-That we men of VFP will lovingly support Women, especially BIPOC and LGBTQ2S Women, in leading our organization out of white supremacy patriarchy to true equity and justice for all

and lastly

-That we men will choose to have fun, <u>lots of fun</u> – without hurting ourselves or others

Roberto Schiraldi

(This poem was originally part of a panel discussion on preventing sexual assault)

*(From my first book, "Healing Love Poems for *white supremacy culture*", 2nd edition, 2021).

(Since we are the ones primarily responsible for perpetrating the vast majority of violence in the U.S., since the beginning)

Two-Spirit*

Walking a spiritual path of healing, light and love, embracing and embodying a deep connection with the universal tones and energies of the Sacred Feminine and the Sacred Masculine, all in one body, one mind, one spirit.

Sacred Feminine	Sacred Masculine
Mother Earth	Father Sky
Place of:	Place of:
nurturance	warrior
gentleness	strength
kindness	courage
forgiveness	perseverance
comfort	protection
compassion	impeccable honesty
humility	dependability
acceptance	equality/equity

*This is what I learned from my teachers during my preparation for and participation in many healing ceremonies of the Sicangu, Lakota, on the Rosebud Reservation in Little White River, South Dakota. Most of our teaching came from Tom Balistrieri who was an apprentice to Joe Eagle Elk, a very revered and respected Iyeska (teacher, healer, medicine man) on the Rosebud. While many Lakota, understandably, do not want whites to participate in their sacred ceremonies, Joe Eagle Elk felt that it was important for whites to learn the traditions (as long as done very respectfully), if real healing was to happen. Our tiospaye/family group of college counselors from around the country, was invited by Harold Whitehorse, a Sicangu Chief to attend the sacred Sundance ceremony because we had been trained very carefully about how to respect the Lakota spiritual traditions. For eight years I was blessed to attend the sacred Sundance ceremony, as well as completed my four year hanblecya (vision quest) and agreed to a life long commitment to be a pipe carrier (using the pipe I was given by Harold Whitehorse, to do healing ceremony).

Amongst the Sicangu, the two-spirit hold a place of honor, respected as powerful, highly intuitive healers because of their deep connection with the mother earth and father sky. Depending on which tribe they were part of, two-spirit were often ostracized and persecuted, even killed. I relate this to witches being burned because of their powerful connection to nature. We had special ceremonies to honor those who chose to claim our two-spirit and for our two-spirit sisters and brothers, who have been killed and suffer because of who they are. A small group of our tiospaye claimed our two-spirit in special ceremony.

I believe that at our core, we are all two-spirit. Unfortunately due to fear, we are taught to choose between our feminine and masculine energies. While two-spirit is not necessarily about sexual orientation, expressing our healthy feminine and masculine sexual energy and passion is certainly an important part of healing, joyful, harmonious life. I believe that racism, homophobia, and all the "isms" would be healed if we each embraced our two-spirit.

Hope this helps.
I am honored to know you.
Mitakuye oyasin (all my relatives),
Francesco Roberto Vincenzo

(In his book "Sisters of the Lost Nation', Nick Medina, a member of the Tunica-Biloxi Tribe of Louisiana, tells a painful story of a Two Spirit youth who journey's through fear, Sacred, Love and vulnerability. With integrity, courage, dignity and grace she inspires her elders by confronting issues of racism, drug addiction, sex trafficking and sexual abuse, violence, murder and much more. These are common problems on many reservations).

P.S. Eileen would always refer to each of us as WoMan...and that is one of the main reasons why the universe brought us together, and that we chose each other....that we both chose to honor the Sacred Feminine and Sacred Masculine parts of ourselves and each other.

The bully balance

So many bullies

racist

sexist

homophobic

classist

nationalist

familial

relational

work place

and on and on and on.....

again....

finding the balance here

is key.

So many opportunities…

to learn and grow from.

When and how

do I stand up

to the bullies?

Sometimes

I need to pause

and breathe in

and envision

a good way

to handle

the situation.

Having faith..

that Higher Power

and

goodness

and

love

are always

in me.

It can be

very difficult.

Especially

when the anger

or fear

rises up.

Thus

stepping away..

and

the loving

soothing,

comforting...

reassuring...

breath.

Sometimes

finding the way

to connect

with the bully...

To hear

and feel

what is real

in their heart

and mind...

Sometimes

choosing

to address it ...

another day.....

another way....

Sometimes....

choosing

to stand

and fight.....

I've rarely
felt good
about
turning away......

wasn't being a man.

And yet.....
maybe
wouldn't have
made it
to 78..
if I hadn't.

And now...
picking battles.....

and doing
my best
to be accepting...
compassionate....
loving......

is way
more important......

however....
the prickles of …..
coward....
fear.....
not standing up
for what's right.....
will always
move me
to action......

the goal being.....
loving....
healing....
mending......
action.

So....
each bully.....
is a gift....
to awaken me ….

to be alert.....
to be fully present.........

to look at the bully .....
deep in their eyes........
in their hearts........
even just for a moment...
in time....

and maybe
that connection....
that recognition
of them....

is really
what they
have been
longing for
all along.....

doesn't mean
it will always work....

for folk's pain
runs deep....

fueling all
the horrific
"isms",
oppression
and trauma.

And
I can be
the bully...
to myself......
to my love ones....

By controlling,

criticizing,

being insensitive,

and unappreciative..

on and on......

So....

I will
continue
to Be Committed

to Be
and Do
My Best

to address
the bullies

inside me ..

and others....

as the universe
brings them
on my path.

~~

May you
choose
to do the same....

and may we all help
each other
on this amazing
racial justice journey.

Lessons From Dad

Like most of us....my dad was far from perfect.
And.... he taught me some very important lessons that framed much of my life. Like being honest, working hard (had me work in construction at an early age, which scared me enough to stay in school)....and standing up to bullies.

He told me how when he was young growing up in Brooklyn, he would confront / fight bullies, who would call him guinea, wop, or dago. He didn't know what the terms meant, but that they were a disrespectful put down. And when we moved out to East Rockaway from Brooklyn, he refused to join a business men's organization because they wouldn't allow men who were Jewish, Puerto Rican or Black. It wasn't til much, much later that I actually learned what those put downs meant....guinea....from the coast of Africa (even though I remember my grandfather telling us that our ancestors came up the boot of Italy from Africa), wop...without papers....dago...still don't know. Either way, I do know the feelings that caused me to confront and fight when I felt put down, hurt, disrespected, less than etc. And that sure has contributed to me doing racial justice work. So...thank you Dad, big time.

P.S. I've had some folks say..."oh, have a sense of humor, you're too sensitive, we didn't mean anything by it".

Yeah, that may be the case...however. I don't feel that putting each other down, like so many of us guys do, especially when we're younger, is funny. I feel like it's ignorant, disrespectful and hurtful. And I have a great sense of humor. I have engaged in gentle ribbing, but have learned to be very careful with it. That's probably another whole book. Bottom line for me...if it contributes to someone feeling uncomfortable, disrespected, or put down.....don't do it.

P.S.S. Many years later, I got brave enough to confront my dad on how scary he was, and how much fear he caused me. And while he initially got defensive, to his great credit, he listened and actually apologized. I will always Love You Big Time for that Dad.

Assimilation vs. Acculturation

A program offering from some years ago:

"Assimilation vs. Acculturation.....Oh the myths we bought into!"

Assimilation means to "adopt" the norms of the "dominant" culture, often at the expense of our unique backgrounds and traditions.

Acculturation means to "adapt to" the norms of the "dominant" culture, while still honoring our unique backgrounds and traditions.

Together we will explore how the "dream" of the melting pot and related pressure to conform, prevents us from real connectedness.

Discussion Questions:

Why is is easier for some ethnic groups to assimilate than others?
What is your background, and what did you miss out on because you and/or your family, assimilated?
What are your feelings about that?
What can you do about that now.

Some additional descriptions:

Assimilation means the absorption, usually total, of an individual or groups of individuals into a dominant culture, often as the result of immigration or in the bad old days, conquest. Their cultural behaviors will conform to dominant norms and previous cultural behaviors will fade, particularly generation to generation.

Acculturation is a less familiar word, but will generally be understood as meaning the adoption of specific cultural behaviors or practices from one culture into another. This does not require but may be the result of population migration. For example, yoga can be considered as an adopted cultural practice in North America.

In sum - In assimilation, people adopt the new culture and lose their original features. A one way process. In acculturation, people retain their original cultural features while adapting to the new culture. A 2 way process.

~~~~~~~~~~

What assimilation brings up for me:

-Indian Boarding Schools- children ripped from their families and culture (often to never return), forced to cut their hair, wear stiff white clothing, and severely punished (sometimes killed) for speaking their native language. "Kill the Indian, save the man" was a govt. slogan that justified this horrific treatment.

-Assimilation is code for **"become white" or else/ forced homogenization/ "melting pot"**, more specifically, adopt the values of this culture or "go home" / back to where you came from"(even though
the individual may have been born here).

-Walk, talk, dress, "Normally", like we do (wealthy, white, hetero males)

-"USA, USA, USA", we're #1/extreme nationalism, unpatriotic if protest ie. Colin Kapernick, kneeling for national anthem, forced out of football

-extreme fear after 911, people of color felt need to have American flags on their cars and homes to not be targeted as terrorists

-me...Francesco Roberto Vincenzo....(why it angers, sometimes even, enrages me to be called Robert or Bob (all American names...feels insulting and disrespectful), residual frustration with parents not teaching us Italian..."speak good english, dress right, cut your hair", use American name, to "fit it", and not be targeted as "different"...which (though was being caring and protective), made me feel even more different, rather than honoring and celebrating uniqueness. Thus the dilemma with assimilation rather than acculturation.

~

Assimilation Nation

This poem was created and presented in response to a request from a dear friend and racial justice ally for a group of parents and their students, struggling with perfectionism/success pressures, some of whom were suicidal as a result.
It was also presented at a truly incredible, healing gathering called "Coming To The Table", where descendants of slaves and descendants of slave owners come together to heal. This deeply emotional, beautiful, love-filled, yearly gathering was created in response to Dr. Martin Luther King's dream of the descendants of the enslaved sitting down at the table of brotherhood (and sisterhood)with the descendants of enslavers.

Assimilation Nation.....
An alliteration recitation
Offered for some healing
With hope that we all reclaim
who we really are.

The original racial bullying
Assimilation......
a very clever **manipulation**.

The birth of a nation, "assimilation nation",
manifest destiny / ultimate **fabrication**,
liberty and justice for all - yep,
if you're white wealthy male elite...
setting all the immigrants against each other,
cause no one wants to be on the bottom of the
heap...huh...
great melting pot...
thought we were all in this together...
yikes I was duped!!!.

militaristic occupation / colonization
forced immigration/capitalization
sterilization / purification/ **homogenization**
stratification / **parentification**
subordination / **degradation** /
dehumanization / **humiliation**
a premeditation / brutaliz**ation,**
traumatization, attempted **annihilation,**
mass incarceration, emaciation, starvation
a forced migration, to the **reservation** and the
plantation...

the original incarceration..
the continual falsification of all the broken treaties
still to this day, no **compensation or conciliation**.....

what about the emancipation proclamation.....hmmmm
again...nice sounding words and intentions.....another falsification
and intentional miscommunication

the lie of race,
subhuman classification, systematic **indoctrination**,
a faulty declaration...
to keep the wealthy entitled male
in control.

ahhh civilization/ gentrification,
economic misappropriation / intoxication,
habitualization /**environmental devastation**....
perfectionistic multi-tasking mechanization........
materialistic **accumulation**.......and more, and more and more......

and we wonder why the mass protestation
and wide-spread altercation,
which leads to mass incarceration
of a whole generation......

and you immigrants and descendants.....
assimilation..just "fit in",

and don't question the boss man,
internment camps...internment camps......
model minority /911/ guantanamo
incarceration.....get the picture?

cut your hair,
walk, talk, behave,
dress to adopt
the boss man's ways
acculturation...
honor your own culture …
in private......
for fear of being accused
of being unpatriotic......

hmmm I thought this country was founded by
immigrants.....
who had to change our names,
deny our language
and our cultures
and play the games
of fitting into ….
the melting pot...
which is extremely hot...
if you are different...
and <u>still</u> so hurtful....

how insulting....
how disrespectful...... ....
dehumanization

fortification / interrogation
perpetual traumatization...
a culture of fear.....

at least it's finally out in the open

and we wonder why
the mass protestation and retaliation....
and resulting incarceration....
with little education and rehabilitation....
on probation, having a record...
can't get a job...
planned recycling back to incarceration
to pad the pockets
of the prison owners and investors.....
ahh capitalization........

....so that's the explanation for the state of this nation.......

And <u>Now</u>......
what's the <u>hope</u> for this land,
when we feel the **alienation**?

If we each do our parts
and we open up our hearts
and we teach our children well
Racial Literacy is what we tell
no more lies or hesitation
just factual presentation
as honest as we can be
the truth <u>will</u> set us free...

So the situation demands
wide-spread **detoxification**,
not the lie of <u>color-blindness</u>,
a disrespectful misrepresentation,
but a strong determination
for a **reidentification**
of the amazing uniqueness
of who we are,
and **courageous collaboration**
from this whole nation
to replace the misinformation,
with the honest presentation,
of what the <u>real</u> values are
that founded this nation.

And this **liberation**
which will finally free us All
for **re-humanization**....

and celebration,
imagine the elation
and the anticipation
for the <u>realization</u>,
inspiration,
amplification
of a beautiful new re-creation
based on real equality,
liberty and justice for alllll y'all......
a real emancipation
and fierce reclamation
of our <u>glorious</u> backgrounds
and histories,
honoring our individual cultures

real unification,
not false homogenization........

…....and then finally.....
we can all come back to the beginning.....
returning home to our beautiful hearts...
a **rememorization**
of who we truly are....t
he greatest power of all.....
LOVE.......

the question is....
Will we be courageous enough
to teach,
and model for our children,
the importance of
a loving foundation,
to be vulnerable and honest.....
for the inspiration
of a new generation....

and hold ourselves,
and each other, accountable,
to the ultimate value,
the most important,
essential ingredient,
to coming back
to our Grand Central Station
of Self- Actualization
and Spiritualization…..

luv...y'all.....**LOVE?**

We've tried everything else...

we have nothing to lose...
and everything to gain.......

so let's buckle up,

and let the joy
and dancing begin.

~

USA...USA

USA....USA

While the
power brokers
play..

Make the
power brokers
pay...

Make the
power brokers
pay.

For All
we pray

Fair values
lead
the way.

Only then
can we
finally
say...

This
is
our
finest
day.

white supremacy as Addiction / 12 Step Recovery

White supremacy can be viewed as an addiction, or powerful habit. This deeply ingrained habit injures the oppressed (as well as the oppressor) because it dehumanizes both parties. Continual denial about our internalized racial superiority often leads to habitual behaviors which are hurtful and feed inequity. For example, we may have automatic visceral reactions when we encounter someone of a different color. When we harbor unconscious biases towards others it is very difficult, if not impossible, to treat them fairly, compassionately or humanely.

When we have not been taught to carefully examine the internalized superiority of ourselves and our society, we will continue to perpetuate racism, even though we might say that we don't want to. It is a habit.

The Good News......like all habits, it can be broken.

The Twelve Steps of Alcoholics Anonymous help to bring sobriety, balance and harmony, and to save the lives of millions of individuals. Applying Twelve Step principles can provide a helpful guide to addressing white supremacy. With brave, humble, and compassionate commitment to life long growth, we can free ourselves from this addiction, and discover the life of mutual love that we all deserve.

Twelve Step Recovery from white supremacy
(Inspired by the work of Dr. Gail Golden, EdD LCSW and others)

1. We admit we were powerless over our white supremacy conditioning – and that the lives of People of Color (and our lives), have been made untenable as a result.

2. We have come to believe that doing our own inner work will restore us to sanity - and that freedom and balance can result

from honesty about our history and our complicity. We cannot do this work alone, so we made a decision to seek leadership from People of Color.

3. We have made a decision to commit to life long learning and teaching about accurate history, and to uncover every vestige of white supremacy that lives inside us (especially the effects of "gendered racism"/ white male supremacy we learned from our founding fathers, and the harmful foundational values exemplified by the Doctrine of Discovery by god's anointed ones, Manifest Destiny and the Constitution never intended for "all the people" ie., Indigenous People and Women excluded and Black People counted as 3/5 of a person, which all contribute to unconscious white superiority).

4. We make a searching and fearless moral inventory of ourselves, and how we are complicit with white supremacy policies and practices.

5. We admit to ourselves and another human being the exact nature of our wrongs.

6. We are entirely ready to learn from others who have effectively done this work.

7. We humbly ask for help from allies and from People of Color, (especially Women of Color) who are willing to lead and guide us– being clear it is <u>our responsibility</u> to do our work and to always hold ourselves accountable, and be accountable to people of color.

8. We make a list of persons we had harmed, and became willing to make amends to them all,

9. We make direct amends wherever possible, except when to do so would injure them or others – and only after clear understanding about our behavior, and well thought out behavior changes and plans so as not to repeat the behavior.

10. We continue to take personal inventory and when we were wrong promptly admitted it (the truth will set us free).

11. We commit to continue to improve ourselves through ever increasing our awareness of white supremacy

12. Having experienced awakening and relief as a result of these steps, we commit to carrying this message to all other white people, and to seek opportunities to promote these principles in all our affairs.

With great thanks to Dr. Gail Golden, EdD, LCSW, the People's Institute for Survival and Beyond, and the United Church of Canada, and much love and respect for all of my sisters and brothers in Twelve Step Recovery. The following three resources can be powerful aids in our ongoing recovery from the white supremacy cultural values which plague us all:

"White Privilege as an Addiction", article by Dr.Gail K. Golden, EdD, LCSW, 2011. For more information see: www.antiracistalliance.com and www.goldenwrites.com, Contact Gail Golden at: peacepoet@aol.com .

Undoing Racism Principles, People's Institute for Survival and Beyond, www.pisab.org. (these principles include being accountable, learning, organizing, allieship and being gatekeepers).

"Twelve Steps toward Ending White Supremacy", Blog post as part of a series for the International Day for the Elimination of Racial Discrimination, The United Church of Canada, March 9, March 21, 2021

**None of what follows is issued as a put down.
It is offered as a perspective to consider....
In our efforts at healing.**

It's what I see as the core element, rarely discussed, that just might contain the cause / healing for our challenges with racism, sexism, militarism and on and on.

What If Washington Was Gay?

What would the implications Be

for ourselves...

for the government...

for all institutions.....

for the Veterans for Peace Board...

and for our Stance

against Militarism

and

Racism?

How might our fear of being Gay

be similar to

our fear of being Racist?

How does all above

relate to

wealthy white hetero male supremacy

cultural values?

~

**The Problem
As I See It.**

Being straight
is the way
anti gay
all the way.

Since day one
macho man

suit of armor
rusted can.

Afraid of weak
safety we seek.
protection our task
so we wear a mask.

<u>Alienation from ourself</u>
put our <u>feelings</u> on the shelf....
put our <u>feelings</u> on the shelf
<u>alienatation from ourself.</u>

Only one way out
by going all in
acknowledging our fears
releasing our tears.

Honoring our feelings
finally bridges the gap
being true selves
we finally beat the rap

Of being hard as nails
chasing our tails
Distant at best
Everything a test.

Finally we can breathe
Finally....we....can....breathe
and go back to the start
as we live from our heart.

~ ~ ~

One Brave Man

Since the beginning of time

we men have feared.....

ourselves.....

each other.....

women.....

animals......

Mother Earth........

Why else would we <u>feel the need</u> to lord power over All Things?

And a group of wealthy land owning white hetero? Males...

stiff as boards, wearing fancy ruffles and wigs....

established a government.....on the surface

with liberty and justice for all......

really for themselves.....

Not for Native People, Not for Black folks, Not for ...Women.....

Not for the Animals......Not for the Mother Earth....

Power over all.....but themselves....

Starting out with military might.....

Power ...wealth ....control......core values

success.....competition to be #1 at all costs
no matter who we need to walk over.

So if we are afraid of being vulnerable
and sharing our emotions.....
for fear of being hurt, rejected, taken advantage
of etc.,........
no wonder being "smarter with words" and being "hard"

are so valued.

Which begs the question....

What is <u>real</u> strength?

A Native saying is...
"Nothing is so strong as gentleness
Nothing so gentle as real strength".

If we are afraid of
"Loving" ourselves
and possibly
another man...

No wonder we
set up walls and barriers...
and need
to declare war..
and to fight.

"Make Love
Not War".

If values such as
compassion, kindness, sharing, equity, fairness, respect,
commitment, gentleness, impeccable integrity, trust,....
were our core foundation

that we commit to living by on all our decisions...
.in all our relations.....in all our institutions......

racism, sexism, homophobia, xenophobia, climate crisis
pandemics...and on and on....
would cease to rule the day..
and love would guide our way.

~

If one brave man
especially one
wealthy white hetero white man
would stand up and say
"Our values are flawed...
from day one".

I acknowledge my part
and urge you
all my brothers
to do the same...
and to give up
the tenuous hold
on core flawed values
that are hurting/killing
us all.

Then we could
love ourselves
and each other
without fear.......

that we might be
weak...
or gay
or feminine
or trans
or whatever......we are so afraid of....

and then we could finally
cease our posturing
behind a wall of words and actions
which alienate ourselves
from ourselves
and each other.

What say you Mr. President?

What say you Congressmen?

What say you Institutional Heads?

What say you Educational Heads?

What say you Religious Heads?

Will one brave wealthy white hetero?male finally stand up and speak this truth?
So we might All be free to finally choose to live by healthy human values...once and for **All**

I for one would back you.

And I know there are many others

Who would join this Loving Values Re-Creation.

GAY   STRAIGHT   CONTINUUM

FLAMING HETERO   FLAMING GAY
ON ONE SIDE ///////////// ON THE OTHER

    MOST OF US......SOMEWHERE IN BETWEEN.

    SO WHAT'S THE BIG DEAL?

AND FOR THOSE OF US
WHO LIVE ON ONE END OF THE SPECTRUM
OR THE OTHER.....

COOL!

AS LONG AS WE DON'T
SIT IN JUDGEMENT
OR IMPOSE OURSELVES
ON ANYONE ELSE.

EACH OF US
FREE TO BE
OUR TRUEST SELVES..
WITHOUT FEAR OF BEING PUT DOWN.

EACH A VALUABLE PART
OF THE
BEAUTIFUL TAPESTRY OF LIFE.

Initial Strategies For Engaging in Racial Justice Work

The following are offered as some preliminary suggestions when considering to do this very difficult, yet extremely rewarding work:

1. Educate ourselves through reading. Since we have not been taught this information it is our responsibility to continue to learn all we can. There are some initial essential books like, "The Fire Next Time" by James Baldwin, "How to be an Antiracist" by Ibram X. Kendi, "Between the World and Me", by Ta-Nehisi Coates, and "Tears We Cannot Stop" by Michael Eric Dyson. Then there is an ever growing body of literature, articles and books such as, "No I Won't Stop Saying White Supremacy" (article), "White Fragility" by Robin DiAngelo (book), "Me and White Supremacy" by Layla Saad (book), "Feeling White", by Cheryl Matias (book). "Waking Up White" by Debby Irving (book).
2. Educate ourselves through training. There are numerous training opportunities, didactic interactive, in person and on line. Some wonderful trainers and training organizations are:

Beyond Diversity Resource Center, beyonddiversity. org, Dr. Amanda Kemp (dramandakemp.com), Dr. Nathalie Edmond (drnatedmond@mmcounselingcenter. com, White Awake,(whiteawake.org), Peoples Institute

For Survival and Beyond (pisab.org), Not In Our Town Princeton (niotprinceton.org). Training for Change (trainingforchange.org).

3. Do our best to integrate all information and experience. Am I willing to keep peeling away the layers for the rest of my life? How do the information and experiences touch me emotionally, mentally, physically, spiritually, socially, and how do I care for yourself well in all those aspects of myself. This is essential...as the work can be very trying, to say the least. Pacing is huge. Engaging in counseling support with a specialist in multicultural counseling can be extremely helpful. And celebrating small accomplishments is huge. Consider making an ongoing commitment to do this work (see below).

4. Search for local ally groups, and do our best to establish trusting relationships with at least a few like minded individuals. This is also essential to have a supportive, sounding board to be vulnerable with and brainstorm and practice strategies. Consider supporting local Black Lives Matter groups.

5. Then, when we decide we are ready....choose to engage in "broaching"..ie., addressing offensive behavior, always guided by first holding ourself in the place of unconditional love and acceptance, and then holding the other in the place of unconditional love and acceptance. And being very gentle and kind with ourselves for being brave enough and caring enough to do this work, regardless of the outcome. Again... this is messy work, we will not do it perfectly, we keep on doing our best, with integrity, compassion and courage.

Racial Justice Pledge

I pledge to continually learn about how this country was founded by and continues to be run by white supremacy values, such as power, wealth, control and elitism.

I pledge to continually remember how white supremacy cultural values hurts me, and our common humanity, and prevents true unity, equity and justice for all.

I pledge to continue to do my best to uncover, heal and change every vestige of white supremacy culture in me, and to support others in doing the same.

I will do my best to live with and share loving kindness and integrity in every thought, word and action.

I will seek and welcome every opportunity to work for unity, equity and justice for all.

To Rosa Parks. An Inspiration in Courage, Perseverance, Patience, Sacred and Love. Thought of this after buying your stamps. Thank you for offering your life. That we may walk in grace and beauty.

Rosa Parks

Ahh the **patience**
I'm learning

A core essence
of my growth.

And my
white
heterosexual
male
supremacy-laced
privilege.

And the postage stamps
of Rosa Parks,
help me to re-member.

When
I forget....
my **patience.**

And my passion
leads me to gallop..

when slow breaths
and gentle words of kindness
are what's needed.

Harriet Tubman comes to my heart/mind, as yet another great example for us all, of the incredible Courage to walk through Fear, with Sacred, Love, Strength and Vulnerability.

These remarkable women, are just two examples of the millions of women of color, immigrants and refugees whose daily lives provide truly inspirational stories in courage and Love

Thank you Aminata, for re-minding me of the importance of patience.

The Mother Love Creator, Father Love Creator, Great Spirit, Wakantanka, Tunkasila.... The Universal One, never fails to re-mind me of the sacred, loving path, and the ongoing work to restore balance.

A Final Racial Healing Story:

-My beloved friend and brother John invited Eileen and me to accompany him and his beloved Liza, in support of some distinguished Lakota Elders who were part of a delegation traveling east to retrieve the remains of Lakota children who died at one of the first Indian boarding schools. The children had been forced to leave their families and tribes (to be "assimilated", "kill the Indian, save the man"), often treated horrifically, many never seeing their loved ones again. Their families and tribes had been previously prevented from taking their remains home. Now, over a hundred years later, they can be brought home. The wounds so deep...so deep. Yet, finally, maybe now, their spirits can be free. And a little healing can start.

For a moment....if you would.....please take a deep breath...close your eyes... and try to imagine how you would feel... in your heart and body.... if one of these children.... is your child.

I honor you three elders for your dignity and grace....
to you, soft spoken, fierce protector grandmother
of the children..... to you, truth speaking, holding
to the fire warrior..... and to you, spiritual guardian
bringing beautiful wisdom, and loving joy to balance
the suffering... Kola. (To respect your privacy and the
dignity of your grieving process I am omitting your
names). Words can barely touch the feelings. You each
in your precious way, teach me about the importance of
courage, integrity, perseverance, compassion, humility,
generosity...... ....forgiveness. Your example, your lessons
will live on in my heart, my mind and my spirit. Wopila.

The Children

Oh... the Children.
their spirits
yearning to be free.

Finally they can return
home again
to their loved ones.

And their spirits
can continue
on their sacred journey.

Honoring Our Children
…..from the plantations
…..from the boarding schools
…..in the border detention centers
…..in human trafficking
All of our children, everywhere
For seven generations to come.

All children want to know..
Am I loved?
Will I be comforted?
Is the world safe?
The answers determine
a kinder world.

In saving our children,
We save our country.

In saving our children,
We save our world.

In saving our children,
We save...ourselves.

Honoring the children
through loving action
is how we can re-create
a loving world.

Mitakuye Oyasin
All My Relatives

Please keep the spirits of the young ones, who have suffered so much, in your loving hearts and prayers ..... their families and loved ones, …. the elders who continue this difficult healing work.... and all those who work to make it right.

~

"When the first chakra is disconnected from the feminine Earth, we can feel orphaned and motherless. We look for security from material things. Individuality prevails over relationships, and selfish drives triumph over family, social and global responsibility. The more separated we become from the Earth, the more hostile we become to the feminine. We disown our passion, our creativity, and our sexuality. Eventually, the Earth itself becomes a baneful place. I remember being told by a medicine woman in the Amazon, "Do you know why they are really cutting down the rain forest? Because it is wet and dark and tangled and feminine"
- Alberto Villoldo, Ph.D,
Dance of the Four Winds: Secrets of the Inca Medicine Wheel.
(With loving appreciation, once again, to my dear friend and colleague Dr. Maria del Carmen Rodriguez, for sending the above quote to me, and for her loving help with editing much of this book.)

~

~

"I am going to venture that the man who sat on the ground in his tipi meditating on life and its meaning, accepting the kinship of all creatures, and acknowledging unity with the universe of things, was infusing into his being the true essence of civilization."

**Luther Standing Bear (1868?-1939)
Oglala Lakota chief.**

~

*"You might say I'm a dreamer,
 but I'm not the only one.
I hope some day you'll join us,
 And the world will be as one"
    -"Imagine"
    John Lennon*

Racial Justice Blessing

May we be instruments of healing and peace...

May we feel discomfort with easy answers,
half-truths, and superficial relationships....
so that we will live deep within our hearts.

May we feel anger at injustice,
oppression and exploitation of people.....
so that we will work for justice, equity and peace.

May we shed tears for those
who suffer from pain, rejection, starvation and war....
so that we will reach out our hands to comfort them
and to soothe their wounds.

May we have clarity of vision
to think that we can make a difference in the world....
so that we will do the things which others
tell us cannot be done.

And may we speak the truth....
and act with courage, and kindness
to honor our ancestors and our children....
so we All will be Free.

Ah Ho. Let It Be So,
Mitakuye Oyasin (all my relations / all my relatives /we are all one).

(Adapted With gratitude from Give Us This Day)

Healing Meditation

Breathing in
I am peaceful and calm

Breathing out
I release all worry and fear

~

Breathing in
I feel safe and protected

Breathing out
I release all worry or fear

~

Breathing in
I am soothed and comforted

Breathing out
I release all sense of urgency

~

Breathing in
I am nourished and nurtured

Breathing out
I let go of outcomes

~

Breathing in
I am filled with Gentle Love

Breathing out
I release all pain and suffering

~

Breathing in
I feel one with all

Breathing out
I send love to all

~

Breathing in
I Am Grateful.

Breathing out
I give gratitude to All.

Two Final Eileen Stories

Eileen was Coordinator of the Community Gardens Program in Phily. for thirty three years. She worked with the Black and Brown communities, helping to turn abandoned lots into beautiful gardens which included growing sustainable food. They were places to gather and sit, and to have gatherings that celebrated their cultural traditions and food. She loved her people and felt they were all family. She was so sad when she had to stop working due to her health. However she felt so blessed for all her loved ones.

She told me a story one time that many didn't know. On one of her first days she was approaching a home to meet with one of the block captains to discuss plans for a new garden. Some gang members thought she was a narc, and beat her up pretty good. Even though she had to go to the hospital with a concussion, she was back on the job the next week. And that local community made sure she was protected from then on.

Fear, Sacred, Love, Vulnerability.

While this last one is not about racism specifically….it is about about Fear, Vulnerability, Sacred, Love…..and the profound connection to Mother Earth and All Things. And it is again about my Great Love Eileen who was and is my main support and ally on this racial justice journey. Many times when I was/am tired and discouraged, she would hold me, and encourage me to to the Loving thing. We made one last trip to the ocean before she passed. It was night time and we could see a storm coming as we stood on our balcony. We could hear the thunder, see the lightening, and feel the wind (It reminded me of a time on the reservation when I was honoring one of my spiritual commitments to stay up on a hill at night, when a magnificent and scary lightening and thunder storm came over me. Yet, from the teachings I'd been given, I knew all the Spirits, The Thunder Beings, The Trees, The Animals, Earth Mother...were all protecting me, and were there to teach me....I saw them, I felt them ....I was safe, and learned many lessons that night).

So Eileen agreed to come down to the beach with me, even though she was scared. Fear, Sacred, Love, Vulnerability....
I reassured her we'd be safe and she trusted. We laid down on the sand, holding hands, breathing.....and taking in the amazing show for the ages.........and the incredible Thunder Beings, Lightening Spirits, Wind Spirits, and Rain Spirits passed right over and around us.....with only a few sprinkles of rain to smile on us.

Crazy? Maybe.

Fear....Sacred.....Love......Vulnerability.

Definitely.

All One...All Connected, All My Relations, All My Relatives. All Connected, All One.

Epilogues

"We The People was never intended to mean
All the People.

The Constitution excluded Women, Native People,
and counted Black Men as 3/5ths a person,
thus protecting the interests of White land owning men."
Mark Charles, Navaho, Dutch
'We The People' / Ted Talks

"Nothing can be changed, until it is
faced (or acknowledged)."
James Baldwin

"Be Impeccable With Your Word"
Don Carlos Ruiz
'The Four Agreements'

***"The Master's Tools Will Never
Dismantle The Master's House.*****
They may allow us temporarily to
beat him at his own game,

but they will never enable us to bring about genuine change."
Audre Lorde
*(For those interested in **real** change, please see this powerful, little gem of a book by the same name).

Dear Kindred,

Thank you for choosing to delve into this journey to the heart and soul and spirit of us men....so we can heal this dis-ease of racism, and all the related pain in our world.

May you honor yourself for being kind, humble, brave and honest, as you walk through the fear to be vulnerable.

And may you choose to honor Yourself and All life....as Sacred, and Loving.

With Much Love and Respect,
Roberto

Gratitude To The Max

To You Who I Call On...Without Thinking....With Definitely Feeling....For Honoring Me With Your Loving, Acceptance, Support, Encouragement... For Supporting Me In Walking Through My Fear....To Vulnerable Loving Action:

Inspira Joy...

.........Johnny...

.............Maria del Carmen...

..................Robin...

......................Aminata...

...........................Allistair...

.................................Andrew...

......................................Emily...

..Phil...

...Ann...

..Rags...

...Mom.

Whew!...That's Some Real Love...Right Here.

Mother Love Creator
     Father Love Creator
Great Spirit
     Universal One

Thank You For Filling My Heart With Your Love
My Mind With Your Wisdom
My Spirit With Your Spirit

May This Offering Touch The
Hearts, Minds and Spirits
Of Those Who Choose to Venture On This Path
Through, Fear, Sacred, Love, Vulnerability

And May We May Each Choose Gentle Love
For Ourselves
And For All Our Relations
For All Life....All Ways.

Pilamaya, Gracias, Mille Grazie

Additional Recommendations

'The Wisdom of Trauma': A Journey To the Root of Human Pain, and The Source of Healing', Dr. Gabor Mate, (documentary, healing from the severest trauma of being alienated from ourselves, which contributes greatly to all the other traumas).

'Say The Wrong Thing', Dr. Amanda Kemp, (helps us be brave to take a Loving action).

'The Wisdom Walk to Self-Mastery: Ancient Wisdom for Transforming Pain', JojopahMaria Nsoroma (how to choose Love over Fear).

'The Power of Vulnerability', Brene Brown, Ted Talks. (the Loving gift of vulnerability)

'A Male Guide to Women's Liberation, Gene Marine (frees us men from the binds of sexism)

'The Four Agreements', Don Miguel Ruiz (being impeccable with our word is the key)

'The Macho Paradox':Why Some Men Hurt Women and How All Men Can Help. Jackson Katz (frees us men from the binds of sexism)

'Leading Men: Presidential Campaigns and the Politics of Manhood', Jackson Katz, **(politician debates are a disgusting example of two bullies in a school yard trying to out duel each other, by putting each other**

down, so disrespectful, and this is the model for our youth, of the leaders of our country)

"Conversations With God", Neale Donald Walsch, (series of books that show us a God who says, "Go inside and you will know, by asking "What would Love do now?").

**'We The People', Ted Talks, Mark Charles, Navaho/ Dutch ("We The People was never intended to mean All the people").

Roberto Schiraldi

Dr. Roberto Schiraldi, EdD, LCP, LCADC is a licensed professional counselor, licensed clinical alcohol and other drug counselor, and has been a racial justice advocate, trainer, and trauma therapist for over 40 years. Roberto is retired from Counseling and Psychological Services at Princeton University, where he was coordinator of the alcohol and other drug treatment team, and was previously employed in a similar capacity by Temple University, where he received his doctorate in Holistic Health Education and Counseling. He is a past President of the New Jersey Association for Multicultural Counseling, past Co-Chair of the Ethics Committee of the New Jersey Counseling Association, and has been a member of numerous racial justice organizations, boards, and committees. He is a pipe carrier in the Sicangu Lakota Native Spiritual Healing Tradition, a Vietnam era veteran, and member of Veterans For Peace.

Roberto has authored two other books in the Racial Justice series: 'Healing Love Poems for white supremacy culture: Living Our Values', and 'Unexpurgated* Racial Justice Poetry with Healing Meditations.

~

To contact me, please go to my website www.robertoschiraldi.com, with related podcast interviews and information about how to order my other two books.

Printed in the United States
by Baker & Taylor Publisher Services